TOXICOLOGY
Principles and Methods

SECOND REVISED EDITION

TOXICOLOGY
Principles and Methods

SECOND REVISED EDITION

M A SUBRAMANIAN
Principal
Sree Amman Arts and Science College
Chittode, Erode
Tamil Nadu

MJP PUBLISHERS

First Edition: 2004

Cataloguing-in-Publication   Data

Subramanian, M.A. (1949 -)
     Toxicology: Principles and Methods / by
M.A. Subramanian. - 2nd ed., rev. -
Chennai : MJP Publishers, 2010
     xxiv, 326p. ; 21 cm.
     Includes Appendix, glossary, references and index.
     ISBN 978-81-8094-078-1 (pbk.)
     1. Poisons and poisoning 2. Plants effects
of poisons on, Molecules I. Title
     615.9 SUB MJP 0070

ISBN 978-81-8094-078-1
© Publishers, 2010
All rights reserved
Printed and bound in India

MJP PUBLISHERS
47, Nallathambi Street
Triplicane
Chennai 600 005

Publisher : J.C. Pillai
Managing Editor : C. Sajeesh Kumar
Marketing Manager : S.Y. Sekar
Project Editor : P. Parvath Radha
Acquisitions Editor : C. Janarthanan
Editorial Team : B. Ramalakshmi, N. Yamuna Devi,
M. Gnanasoundari, Lissy John, R. Magesh
CIP Data : Prof. K. Hariharan, Librarian
RKM Vivekananda College, Chennai.

This book has been published in good faith that the work of the author is original. All efforts have been taken to make the material error-free. However, the author and publisher disclaim responsibility for any inadvertent errors.

PREFACE

TO THE SECOND EDITION

I take great pleasure in presenting the second edition of the book *Toxicology: Principles and Methods*. Since its publication, the first edition of the book has received overwhelming enthusiasm and responses in the form of constructive criticisms and positive suggestions from the academics. Therefore, an additional effort for a comprehensive and thorough revision has been made in this edition with more information in each and every chapter. The chapter "Toxicity at the level of organs and organ systems" has been completely rewritten and a few new chapters have been included—"Persistence and residual nature of toxicants," "Basics of nutritional toxicology", Liver function tests in toxicology and "Antidotal procedures"—along with additions in appendix, index and glossary.

There is no doubt that this revised enlarged edition of the book would be highly informative and useful to the student and teaching community in the field of toxicology.

The author records his appreciation and thanks to MJP Publishers, Chennai, for their participation in making the first edition of the book a remarkable success.

M.A. Subramanian

PREFACE

TO THE FIRST EDITION

The ever-increasing human population, industrialization, technological advancements and urbanization have resulted in the degradation of the environment through the release of large quantum of a variety of chemicals. A toxicant is one which causes adverse effects on biological systems, retarding physiological and biochemical functions thereby leading to death. Therefore, it is of paramount importance to detect these substances in the ecosystem and to evaluate the harmful effects of these xenobiotic chemicals on life. Toxicology is the study of the route of entry of toxicants, their absorption, distribution, elimination and metabolic transformations, their mode of action and persistence, target-site interactions, impact assessments and their dynamics in organisms and environment. The ultimate aim of toxicology is to protect life especially humans against the toxic substances.

This book emphasizes on the main principles of toxicology and contains a few toxicological methods in order to provide a basic background to those who enter anew into this field. Efforts have been taken to present the contents in a simple manner.

I gratefully acknowledge a number of standard books and research articles which have provided valuable information for writing this book. The information provided in the chapter "Impact of pollutants on aquatic organisms" has been obtained from observations made at our college laboratory. I am thankful to my family members and friends for their constant help and encouragement. I am also thankful to MJP Publishers, Chennai, for publishing the book on time. Constructive criticism and suggestions for the improvement of the book will be greatly appreciated.

M.A. Subramanian

CONTENTS

1

INTRODUCTION

Toxicology literally means "study of poisons". The word is derived from the Greek words *toxicon* meaning arrow or poison and *logos* meaning study. This branch is typically a multidisciplinary field of science which deals with the harmful effects of toxic substances on individual organisms and involves the analyses of internal distribution of toxicants in the body of organisms and their mode of action. It also deals with the transformation of toxicants within the organisms and the formation of their metabolites.

The knowledge of poisonous substances which occur commonly in nature, has been confronted by man from early times. The ancient Egyptians (1550 BC) had extensive knowledge about the toxic and curative properties of poisonous substances of natural origin. The Greeks and Romans also had a good deal of interest in poisons. Dating from 1673 BC early European literature has plenty of information on toxins. During the 16th century, Paracellsus (1493–1541), a famous physician introduced the "dose principle" of the toxic substances. Bonaventura Orfila is considered as the founder of "modern toxicology" as he laid the foundation for it during the early 19th century. He tested toxins on animals and developed methods for chemical analyses of poisons in tissues and body fluids of organisms. During this period, a number of experiments were carried out to elucidate the harmful effects of many toxic substances. At the end of the 19th century, a lot of manuals were published in journals with the description of the impact of a large number of toxicants and current experimental methods.

After the second world war, new toxicological methods were developed and the first journal on experimental toxicology was published in 1930. From 1960, the realization of the importance of protecting health and the quality of the environment created an increasing interest in the study of toxicology. Today, the field of toxicology is supported by various disciplines such as physiology, pathology, biochemistry, analytical chemistry, organic chemistry, cell biology and molecular sciences.

Scientists who study the toxicology of chemicals and poisoning are known as toxicologists. The main objectives of toxicology include elucidation and research to know the toxic properties of chemicals, evaluation of hazards of chemicals to organisms (risk estimation) and advising the society to control the harmful effects of chemicals (hazard control). In the present era, man is facing a large number of newly developed substances so that a thorough knowledge of the risks caused by toxicants is essential in order to protect man and his environment. To understand the mechanisms and the effects of xenobiotics, all basic medical- biological - and chemical sciences are integrated in toxicological research.

Various subdisciplines like neurotoxicology, immunotoxicology, molecular toxicology, chemical toxicology and genetic toxicology are distinguished depending on the nature of integration between toxicological research and the basic disciplines. The main subdisciplines, which integrate various fields of applied research with the health of humans and environment and provide advice to the society, include clinical toxicology, nutritional toxicology, environmental toxicology and occupational toxicology. Clinical toxicology integrates toxicology, clinical medicine and pharmacy and is concerned with the diagnosis and treatment of poisoning through evaluation methods. Nutritional toxicology integrates nutrition and food science and is concerned with the toxicological aspects of foodstuff and nutritional habits. Occupational toxicology integrates toxicology with occupational medicine and occupational hygiene. Environmental toxicology integrates toxicology with health sciences, ecology and environmental chemistry. Thus toxicology is a multidisciplinary field of science and assumes an interdiscipline branch.

Among the main subdisciplines of toxicology, environmental toxicology or eco-toxicology has been recognized as an important discipline in the recent past (1962) and has attained greater development. This branch is mainly concerned with the study of toxic substances found in the environment and it can be represented as a triangle composed of environmental chemistry, toxicology and ecology. Thereafter, the scope of toxicology has been widely expanded by the amalgamation of all biological systems.

2

TOXIC SUBSTANCES

A **toxic substance** is one which can produce adverse or harmful effects in biological systems by interfering with their structure and function, leading to mortality. These substances are indiscriminately introduced into the ecosystems due to which the environmental quality is impaired. These environmental toxicants are otherwise called as **xenobiotics** (Gr. *Xenon*—a stranger; *bios*—life). From a toxicological point of view, it can be considered that no chemical is either completely safe or harmful. The safe or harmful nature of toxic substances depends on the dosage of the substance and the duration for which the organisms are exposed to it. Poisonous chemicals can produce a harmful impact on organisms when they come in contact with biological membrane systems, in adequately high concentrations.

The fate of a toxic substance in the environment and in the organisms can be conveniently arranged into the following categories.

i. **The exposure phase** which includes the moment at which the toxic substance comes into contact with the organism.

ii. **The toxicokinetic phase** which covers the fate of the. substance in the environment and in the organisms.

iii. **The toxicodynamic phase** which forms the interactions of toxicants with the organism causing harmful effect.

Different phases of the fate of toxic substances are as shown in Figure 2.1.

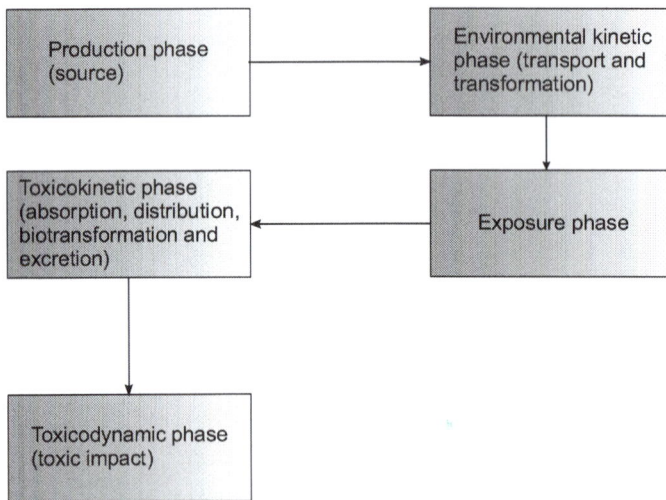

Figure 2.1   Schematic representation of different phases of fate of toxicants

All toxicants belong to two major groups based on their origin, namely **natural toxicants** and **artificial** or **synthetic toxicants**. Natural toxicants originate from animal, plant and mineral sources whereas synthetic toxicants include commercial or industrial synthetic chemicals. In a broad sense, based on the chemical nature, toxicants are classified into **organic toxicants** which mainly include a limited number of substances of natural origin and **inorganic toxicants** which include both natural and synthetic substances.

BRIEF SURVEY OF
ENVIRONMENTAL TOXICANTS

Pesticides

These include a large number of chemicals which are used for the eradication of unwanted pests and are broadly classified into insecticides, herbicides and fungicides.

Insecticides   These are the chemicals used to eliminate insect pests and include **organochlorine** insecticides, **organophosphate** insecticides, **carbamates** and other chemical insecticides.

Organochlorine insecticides   These are chlorinated derivatives of ethane such as DDT, DDD, etc. These are highly stable and are not easily degraded in biological systems. As they are highly persistent, they cause prolonged toxicity and are found to disrupt the transmission of nerve impulses in nerve fibres (neurotoxic).

Organophosphate pesticides   These are the most commonly used pesticides and are easily degraded within the organisms as well as in the environment. The metabolites of these chemicals are harmless and are not accumulated in the body of living organisms. These interfere with the nerve impulse conduction at synaptic regions by inhibiting AchE.

Carbamates   They are also easily degraded in the body of organisms and in the environment and bring about inhibition of AchE as in the case of organophosphates.

Herbicides   These include chemicals such as dinitrophenols, bipyridyls, triazine derivatives, etc. which are used to eliminate unwanted plants. Though most of the herbicides are found to exert a very little toxicity in vertebrates, some are highly toxic to various animal groups including man.

Fungicides   These include dithiocarbamates and many substituted aromatics which are used against fungi. While most of the fungicides exert only less toxicity, the compounds containing mercury are shown to be lethal in higher animals and humans.

Fertilizers

In the present, agricultural practices require a number of synthetic fertilizers to boost the productivity. Though the chemical fertilizers contain a few minerals (nitrogen, phosphorus, potassium, sulphur, etc.) necessary for plant growth, they drain into the aquatic ecosystem through irrigation, rainfall and drainage thereby contaminating it. These nutrients along with organic substances from sewage cause eutrophication in aquatic bodies, causing mortality of fish and other aquatic animals.

Trace Elements and Heavy Metals

Trace elements and heavy metals are non-degradable toxicants which do not degrade but accumulate in the environment causing toxic effects on organisms. In living systems, the **trace metals** interact with metabolically active groups via strong bonds and produce toxic action. **Heavy metals** form a major group of toxicants, since a variety of metals are used in industries and in everyday life of man. The main sources of heavy metals in the environment are industrial wastes, municipal wastes, wastes from power generation plants, and so on. The trace and heavy metals enter into the body of organisms across the biological membrane by active transport. Many of them are capable of forming metallic bonds so that they become highly lipophilic. Toxicologically important trace elements and heavy metals include aluminium, arsenicals, beryllium, cadmium, chromium, copper, cobalt, lead, manganese, mercury, molybdenum, selenium, zinc, nickel, etc.

Radioactive Substances

Radioactive substances are the most toxic and the effect of radiation is tremendously high. The sources of radiation include two categories, namely, **natural** or **background radiation** and **man-made radiation.** In nature, nuclear activity involving subatomic particles produce radiations through a large number of isotopes which are capable of emitting high energy radiations. In addition, radiations such as X-rays, cosmic rays and ultraviolet rays from outer space also constitute natural sources of radiations. The cosmic rays consist of mixtures of corpuscular and electromagnetic components. Radionucleotides such as radium, thorium, uranium and isotopes of potassium (K^{40}) and carbon (C^{14}) are widely distributed in the soil, rocks, air and water. Marine sediments contain higher concentrations of radionucleotides than non-marine sediments. Atmospheric radiations come from radioactive gases such as radon and thoron in the atmosphere. The internal radiation includes minute quantities of uranium, thorium and isotopes of potassium, strontium and carbon which are stored in the body tissues. The sources of man-made radiations include atomic tests, radioactive fallouts, nuclear reactor wastes and use of radioactive elements in industries, medicine and research.

The radioactive substances cannot be influenced by physical, chemical or biological processes and so they are absorbed, accumulated and biomagnified in living organisms. As the radiations have high penetrating power, they are easily absorbed by nucleic acids. The spent fuels and left over materials from nuclear reactors and nuclear fuel-processing units are considered to be **high-level irradiated**

toxicants as they contain a number of radioactive isotopes with a half-life of hundreds of thousand years. The irradiated toxic substances which possess a less quantity of radioactive isotopes with a lesser half-life, are known as **low-level irradiated toxicants.** Some of these isotopes also contain very long half-life. Though the low-level wastes are causatives of little effect, they are as dangerous as high-level wastes. However, both types of wastes are carcinogenic, mutagenic and teratogenic.

disrupt development of embryo/fetus

Food Additives

These are of three types, namely **direct** or **intentional** (flavours, preservatives, emulsifiers, artificial sweeteners, colours, etc.), **indirect** or **incidental** (residues of fertilizers, pesticides, heavy metals, etc.) and **natural** (safrole, aflatoxins, etc.). The indirect food additives originate mainly due to the production, storage and packaging of the products. The natural food additives are produced by processing of materials, metabolic reactions and chemical combinations.

Automobile Emissions

The major toxicants through automobile emissions are carbon monoxide, hydrocarbons (benzene, benzopyrene, methane, etc.), sulphur dioxide, nitrogen oxides, lead compounds and many combustion products.

EFFECTS OF TOXIC SUBSTANCES

The toxicants, after absorption into the body of organisms, are translocated to different parts through body fluids where they interact with biological molecules to form complexes. By enzymatic reactions, they are converted into simple harmless molecules to be excreted. They may also be converted into more toxic metabolites than the parent compounds and produce toxic effects at the site of action. The toxic effects inflicted by different toxic substances are highly variable. The following are a list of effects caused by toxic substances. Some substances cause effect at the region of the contact with the organ concerned (**local effects**). The process of toxic substances affecting the target organs only after being absorbed and distributed are called **systemic effects.** If the toxic effects disappear on ceasing of exposure of toxicants, then they are **reversible effects** in which the toxic chemicals are detoxified and excreted. When the effects persist even after the exposure is stopped, they are **irreversible toxic effects.** The toxic effects, which appear immediately after the exposure of toxicants, are referred to as **immediate effects** whereas the **delayed effects** take some interval between the exposure and the manifestation of the effect. The alterations in the physiology of target organs caused by some toxic substances are termed as **functional effects** which are mostly reversible. The changes in the cross morphology of tissues due to the stress caused by the toxic chemicals are called **morphological effects** which are irreversible. The adverse effects inflicted by the toxic substances in biological molecules are referred to as **biochemical effects.**

ACCUMULATION OF TOXICANTS IN ORGANISMS

The environmental pollution mainly caused by human activity produces a large quantity of toxic wastes which bring about undesirable changes in various components of the environment. These toxic substances make entry into the living organisms from any one of the three media, namely, water, soil and air. The rate of entry of these substances into the biological systems depends on the size of the molecules, concentration in the medium and their solubility in water and lipid. The lipophilic substances are easily carried across the plasma membrane simply by diffusion whereas the hydrophilic substances are mostly transported by active transport. Moreover, smallest molecules can also pass through the pores of the biological membrane. The simple biodegradable substances are easily decomposed but complex biodegradable substances possess a remarkable power of resistance of decomposition and they persist in the environment for a very long period. They enter into the biosphere and concentrate, thus forming potential toxicants to the organisms.

The quantity of a toxicant in an organism represents its balance between intake, conversion and excretion. In an organism if the intake of the substance is greater than excretion and degradation, then it is called **accumulation**. When the average concentration of a substance in an organism is higher than is found in its surrounding environment, then it is called **bioconcentration**. Bioconcentration is influenced by the structures, concentration and water solubility of the substances, physiological status of the organisms, and environmental conditions. In general, bioconcentration of substances (low polarity compounds only) is inversely related to the exposure concentration.

Biomagnification is a process in which the average concentration of a substance in various organisms increases through successive trophic levels in both terrestrial and aquatic animals. Here, the intake of chemicals occurs in a stepwise manner with the toxic residues occurring first in the lower trophic levels and then at higher trophic levels after some delay. Bioaccumulation refers to a high concentration of a substance in an organism after absorption from the ecosystem. This is due to the intake of toxic residues from both water and land and is influenced by the feeding habits of the species and the physical conditions of the environment. In an organism, the intake and accumulation mainly depend on the physiological aspects related to the substance, the possibilities of internal storage, processing mechanism and excretion routes. Because the conversion of many substances in the body of organisms by enzymatic reactions into metabolites are more toxic than the original toxicants. This phenomenon is called **biotransformation.**

TOXIC RESIDUES

The persistence of toxic residues in the soil and their transport to the environment are the major factors which influence the accumulation of toxicants in the aquatic ecosystem. The soil acts as a 'trap' and releases the toxicants to the atmosphere and also to the surface run off. Volatilization, leaching, surface erosion and adsorption to the soil particles are the major factors which influence the mobility of toxic residues in the soil. Many volatile organic compounds readily evaporate from the surface of plants or soil. They may either enter into the atmosphere or slowly deposited by precipitation in the soil as dry particulate matter. Large quantities of toxic residues received by the soil are transported to the water bodies through surface run off and groundwater contamination and pose a threat to the aquatic environment.

The substances which are not water-soluble and are more persistent in soil and aquatic environment. In water, the toxic residues of less persistency will be more in solution than in particulate matter. In water bodies, the residues are partly lost to the atmosphere, partly degraded, partly incorporated in the organisms and partly transported to the sediments.

In the atmosphere, the residues of pesticides and industrial organic chemicals are carried over long distances in association with dust particles. The residues of more persistency will be trapped by the suspended particles of water or by the sediments. In the bottom sediments of the aquatic ecosystem, the residues undergo aerobic as well as anaerobic degradation. The sediments of the ocean are the ultimate reservoirs of the toxic residues. As ocean sediments slowly release these substances, the residual effect may occur for a prolonged period.

Residual Analytical Methods

As the toxic residues occur in extremely low quantities, it is very difficult to analyse them. The following methods are widely used in analysing the toxic residues.

Chromatographic method   After proper sampling, the residues are extracted by using specific solvent systems (hexane–isopropanol, benzene–isopropanol, redistilled acetonitrile, methylene blue, etc.). The cleaning procedure involves either solvent–solvent or solvent–aqueous solution partitioning method (acetonitrile–hexane or acetonitrile–petroleum ether partition) or column absorption method with silica gel, activated charcoal or activated alumina as adsorbent.

Among various chromatographic techniques, gas–liquid chromatography (GLC), thin layer chromatography (TLC) and paper chromatography are widely used. A number of detectors are available to separate toxic residues, the choice of which depends on the selectivity, sensitivity and reliability of operation of the instrument.

Chemical method   Here, the properties of toxic residues are altered with the help of standard chemicals by colorimetric reactions to enhance the detection of the residues.

Spectroscopic method   By using mass spectroscopy, the metabolic processes of residues can be analysed. UV and IR spectroscopic techniques are less sensitive and require complete purification of samples.

Fluoroscopic method   It is used to analyse the residues which are capable of absorbing electromagnetic energy and releasing a part of it as light.

Phosphorimetric method   It is commonly used to study pesticide residues.

NMR method   Nuclear magnetic resonance (NMR) spectroscopy is used to study certain pesticides.

REVIEW QUESTIONS

1.   What are toxicants? How are they classified?
2.   Give a brief account of environmental toxicants.
3.   In what ways are toxicants accumulated in organisms?
4.   What are toxic residues? Explain the methods to analyse them.
5.   Explain the sources of radiation and its impact on the living organisms.
6.   Write an account on generalized effects of toxicants.

3

TOXICITY

Toxicity of a substance refers to its capacity to cause adverse effects on living organisms and the term is more commonly used to compare the impact of two or more substances. Toxic impact may bring physiological, biochemical or pathological alterations in organisms. In other words, toxicity may inflict signs or symptoms of illness varying from simple local effects to complex disorders resulting in the mortality of organisms. The intoxication of toxicants includes a sequence of events which starts with the exposure of a substance to an organism. Subsequently, the toxic substances are absorbed in various routes followed by their distribution within the body of organisms, thus causing an internal exposure.

Within the body of organisms, the substances are converted into metabolites which may either be more toxic *(bioactivation)* or less toxic (detoxication). The whole process of absorption, distribution, transformation and excretion is called **toxicokinetics** of a substance. The studies on absorption, distribution, metabolism and excretion (ADME studies) form the first step in evaluating the fate of the chemicals in the body. The sequence of events, in which the interaction of a substance with target molecules of organisms results in a toxic impact, and is concerned with the mode of action of a substance.

FACTORS AFFECTING TOXICITY OF CHEMICALS

Toxicity of various chemicals is based on a number of factors which are as follows:

Duration of exposure and concentration of chemicals   First, a chemical must come in contact with the receptors of organisms at an adequate concentration and duration in order to react and to cause adverse effects. That is, the concentration of the chemicals and the duration of exposure are the two important factors to have a toxic effect on organisms.

Species   Susceptibility to toxicants varies from species to species. Within a species, size, age, sex, food, physiological conditions, health, etc. influence the toxicity of the chemical. For example, small-sized individuals, immature young ones, unhealthy organisms and females are more prone to toxicants. It is known that protein and carbohydrate diets give protection against toxicity while fat diets make them victims of toxicity.

Environmental factors   Many biotic (population density, competition, etc.) and abiotic (temperature, pH, salinity, dissolved solids, etc. in an aquatic environment) factors affect the toxicity of chemicals.

Nature of toxicants   The toxicity of chemicals is also influenced by the composition and physico-chemical properties of the toxicants as well as their mode of administration and nature of activity (selective or non-selective; synergism or antagonism).

EVALUATION OF TOXICITY

Today, both the developing and developed countries face ecological and toxicological problems due to indiscriminate release of pollutants into the environment. Toxicity is nothing but a chemical's potency to cause an adverse impact on living organisms and is dose- and duration dependent. To evaluate the toxic impact of various pollutants, a number of bioassay procedures have been put to use in the study of concentration levels of poisonous substances on organisms. In other words, toxicity tests are carried out in the laboratory and are useful to assess the toxic impact of the pollutant on living organisms under the standard and reproducible conditions.

In evaluating the toxicity of pollutants, the following terms are used.

Acute toxicity The harmful effect which will bring mortality over a short-term exposure of organisms (up to 96 hrs) to relatively high doses of a substance.

Chronic toxicity The harmful effect which will bring mortality over a long-term exposure of organisms to the pollutants of relatively low doses.

Lethal concentration (LC) The **concentration** of the pollutant which will cause mortality in a particular proportion of the experimental animals. For example, LC_{50} is the median lethal concentration which will kill 50% of treated individuals. The exposure time can also be specified as LC_{50} 24 hrs, LC_{50} 48 hrs, LC_{50} 72 hrs, LC_{50} 96 hrs, and so on. The concentrations above these values are lethal and those below are sublethal.

Lethal dose (LD) The **dose** of the pollutant which will cause mortality in a particular proportion of the experimental animals. The exposure time can also be incorporated as in lethal concentration (LD_{50} 24 hrs, LD_{50} 48 hrs, LD_{50} 72 hrs, LD_{50} 96 hrs, and so on).

Lethal time (LT)   The **time** required to kill the organisms at certain dose or concentration and LT_{50} represents the time required to kill 50% of the organisms.

Effective concentration (EC)   The concentration which will cause a desired effect, usually sublethal in a particular proportion of test animals in which the exposure time can also be specified as EC_{50} 24 hrs, EC_{50} 48 hrs, EC_{50} 72 hrs, EC_{50} 96 hrs and so on.

Effective dose (ED)   The dose of the toxicants which will cause mortality in a particular proportion and the exposure time can also be incorporated as in effective concentration (ED_{50} 24 hrs, ED_{50} 48 hrs, ED_{50} 72 hrs, ED_{50} 96 hrs and so on).

Knockdown dose (KD)   The dose which will cause knockdown of animals in a particular proportion and the exposure time is incorporated as KD_{50} 24 hrs, KD_{50} 48 hrs, KD_{50} 72 hrs, KD_{50} 96 hrs and so on.

Knockdown time (KT)   The time required to cause knockdown of animals in a particular proportion and the exposure time as also be specified as (KT_{50} 24 hrs, KT_{50} 48 hrs, KT_{50} 72 hrs, KT_{50} 96 hrs and so on).

Inhibiting concentration (IC)   The concentration of the toxicant which will cause inhibition of a biological function in a specific percentage.

Medium tolerance limit (TLm)   The concentration at which 50% of test animals survive for a specific exposure time.

No-observed-effect concentration (NOEC)   The highest concentration in which the values of the effect are not statistically significant (in life cycle tests) when compared to the control.

Lowest observed effect concentration (LOEC) The lowest concentration in which the values of the effect are statistically significant (in life cycle tests) when compared to the control.

Maximum allowable toxicant concentration (MATC) The concentration which will not cause significant harmful effects on test organisms.

Static bioassay Toxicity test in which the test organisms and test solutions are kept in chambers for a particular duration.

Renewable bioassay Toxicity test in which the test solutions are renewed periodically with appropriate concentrations.

Flow-through bioassay Toxicity test in which the test solutions are renewed continuously in test chambers throughout the experimental periods.

Additive It is the summation of effects in which different substances will influence each other's action on organisms.

Synergism or potentiation The combined effect of substances is greater than the individual effect of the toxicants. That is, one substance will influence the biotransformation of the other.

Antagonism It is the induction of an effect in which one toxicant will induce the effect of another, in a mixture of pollutants.

TOXICITY TESTS

A large number of toxicity tests have been developed and are classified in the following manner.

Toxicity tests

Based on the effects on sensitive life stages	Based on time duration	Based on addition of test solutions	Based on purpose of the test
	1. Short-term	1. Static	1. Quality
	2. Intermediate	2. Renewal	2. Sensitivity
	3. Long-term	3. Flow-through	

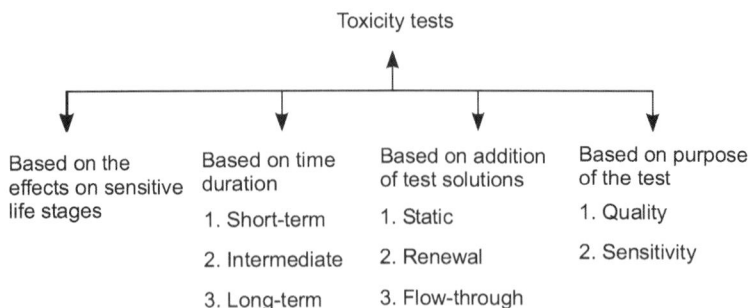

The following are the types of toxicity tests based on the sensitivity.

Life cycle toxicity tests   These are useful to assess the effects of chronic exposure to a pollutant on reproduction, growth, survival, and so on over one or more generations of test organisms.

Most sensitive life toxicity tests   These are useful to assess the effects of chronic exposure to a pollutant on most sensitive life stages such as survival and growth.

Functional toxicity tests   These are useful to assess the effects of pollutants on various physiological functions of organisms.

Depending on the time duration, toxicity tests are classified into the following types.

Short-term toxicity tests or acute toxicity tests   Here, the test animals are exposed to different concentrations of chemicals for a shorter period (24, 48, 72, or 96 hrs) to observe the mortality. These tests are used for obtaining rapid and inexpensive data as well as for routine monitoring of pollutant discharge level. These are also useful to estimate the overall toxicity of the toxicant and to assess the relative toxicity of xenobiotics to selected test organisms. These tests may be static, renewal or flow-through.

Toxicity tests of intermediate duration These test
to evaluate the effects of toxicants on various life stages of
organisms which have a long life cycle. These tests may be
static, renewable or flow-through. Here the exposure period
to the toxicants corresponds to the life stages of the organisms
(a few days) and the criterion for the toxicity is mortality of
organisms or physiological and biochemical effects.

Long-term toxicity tests or chronic toxicity tests These
tests are used to assess the toxicity of toxicants involving
the life cycle of test organisms. The exposure period in these
tests may range from a minimum of 7 days to 4 weeks or
even to several months. The flow-through tests are mostly
recommended with some exceptions. In these studies, the
criterion for the toxicity is mortality of organisms or any
undesirable harmful effects on morphology and physiology
of test organisms. Through chronic toxicity tests, three
important aspects have to be evaluated.

   i. toxicity effects on tissue and organ functions,

   ii. secondary effects such as carcinogenicity,
mutagenicity and teratogenicity and

   iii. no effect level.

Based on the method of adding test solutions, toxicity
tests are classified into the following types.

   i. **Static toxicity test** It is the simplest test in which
the test organisms are placed in containers containing
the test water for 48–96 hrs. The organisms are
removed at the end of the test period and the
mortality of animals are recorded.

   ii. **Renewable toxicity test** It is a more sophisticated
test in which the test solution in the container is
periodically replaced.

iii. **Flow-through toxicity test** Here the diluted test solution is renewed continuously or by periodic additions. Thus this method requires large volume of water for dilution.

In addition, the toxicity tests are classified depending on the purpose of the test as single compound testing (toxicity evaluation of a single toxicant), quality monitoring (assessing the quality of the environment), sensitivity testing (evaluation of sensitivity of organisms to the toxicants), etc.

BIOASSAYS

These are the tests in which the test animals are exposed to pollutants for a specific period at laboratory conditions to observe the impact of toxicants. These tests form the basic criteria for evaluating the toxicants of any pollutant by determining LC_{50} or LD_{50} values. These may be short-term tests or long-term tests. Bioassays may either be static or renewal or continuous flow types and are mainly used in the study of aquatic pollution. The main purpose of carrying out these tests is to assess the hazardous nature of a pollutant to the non-target organisms. Chronic toxicity results in the deposition of chemicals in target sites so that histological, physiological or biochemical alterations can be evidenced in test organisms. They are also useful to assess the discharge of safe concentration of pollutants and to study synergistic effect of chemicals.

Static Bioassay Procedure

Selection of pollutant and diluent medium The pollutant, the toxicity of which is to be determined, is selected and collected from its source in raw condition. In the laboratory, the toxicant is dissolved in the case of a solid, or diluted in the case of liquid by using diluent media to get desired concentrations or doses. In general, the dechlorinated tap water is used as diluent medium in most bioassay methods.

Selection of test organisms In the bioassay tests, the following criteria should be considered to select test organisms.

  i. The organism should be sensitive to the factors to be studied.

  ii. They should be abundant and available throughout the year in a particular geographical location.

  iii. They should possess recreational, economic and ecological importance.

  iv. They should be easily reared in the laboratory and their physiological and nutritional aspects should be known.

  v. The age and quality of laboratory-reared organisms should be known so that the age, life history and existing conditions can be recorded.

Method Bioassays are being carried out in the laboratory by using different types of organisms such as algae, plankton and macroinvertebrates. Until recently, fishes are the most widely used animals to test the toxicity of pollutants in both acute and chronic toxicity tests. In recent times, Western countries insist that the acute and chronic toxicity of dangerous chemicals must be tested on at least three aquatic organisms including a fish. In this respect, the **dragonfly larvae** have proved to be the best test animals in toxicity studies. They are aquatic, commonly available throughout the year and easily reared in the laboratory. They are found to respond to the toxicants at the tissue level so that studies on histomorphological alterations as well as quantification of physiological and biochemical parameters including the activity levels of enzymes are possible.

Feeding the test organisms   There are three options with reference to the feeding of the test organisms in toxicity tests. These include:

Unrestricted food supply   The test organisms can be provided with a continuous supply of food in greater quantities.

Intermittent or satiated food supply   The test organisms can be provided with consumable quantity of food once or twice daily.

Uniformly restricted food supply   The test organisms can be provided with the quantity of food which will be completely consumed.

Determination of LC$_{50}$ 96 hrs Value through Static Bioassay (Acute Toxicity)

The pollutant is made into various concentrations (5 or 6 arbitrary concentrations) by using the diluent medium. A group of laboratory-acclimatized test animals (usually 10 or more) having the same weight and morphometry are introduced into each concentration contained in a test chamber along with respective controls. A minimum of 4 replicates for each concentration are to be kept.

The test animals should not be given food during the entire period of experiment. The experimental media in the test chambers are replaced with appropriate concentrations of the pollutant daily. The mortality of animals in each concentration of the toxicant is observed and recorded daily up to 96 hrs exposure. In each concentration, the dead animals are to be removed from the test chambers.

Table 3.1   Determination of LC$_{50}$ 96 hrs value of a pollutant to the test animals

Concentration of toxicant (%)	No. of animals exposed	No. of animals dead	% Kill	Probit kill	LC$_{50}$
Control	10	0	0	0	
5	10	0	0	0	
10	10	2	20	4.16	
15	10	5	50	5.00	15% concentration of toxicant
20	10	7	70	5.52	
25	10	7	70	5.22	
30	10	10	100	8.09	

The logarithmic value of each concentration of the pollutant is noted. The recorded mortality of animals in each concentration is then converted into mortality percentage and then into probit kill (%) by referring to the table of probit transformation. The results are tabulated in Table 3.1 and graphically represented in Figure 3.1.

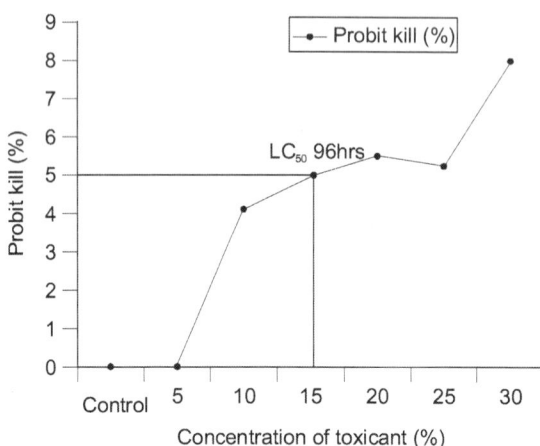

Figure 3.1 Probit kill (%) of test animals at different concentrations of the pollutant after 96 hrs exposure period

EXPOSURE OF ANIMALS TO SUBLETHAL CONCENTRATIONS OF THE POLLUTANT (CHRONIC TOXICITY)

After the determination of 96 hrs LC_{50} value of the pollutant for the test organisms, five different sublethal concentrations (usually around 10% of the LC_{50} value) are chosen for assessing long-term sublethal toxicity of the pollutant. Six experimental containers are setup, each containing ten test animals. The first five containers contain different test solutions and the sixth one is kept as control. Feeding the animals is continued daily and the test solutions are replaced with appropriate concentrations of the pollutant in each container. After the experimental exposure, the animals are used for histological, physiological and biochemical studies.

REVIEW QUESTIONS

1. What is toxicity? Explain the factors which affect the toxicity of chemicals?

2. Differentiate the following:
   i. Acute and chronic toxicity
   ii. Lethal concentration and lethal dose
   iii. Effective dose and knockdown dose
   iv. NOEC and LOEC, IC and MATC

3. How are toxicity tests classified?

4. Explain the criteria for selecting and rearing the test organisms in bioassays.

5. Describe the procedure to determine LC_{50} 96 hrs value of a pollutant.

4

ROUTES OF EXPOSURE OF TOXICANTS

In qualitative terms, exposure is defined as the condition in which the xenobiotics come in contact with an organism by penetration. The harmful effects of the toxic compounds on organisms mainly depend on the form in which the organisms are exposed. The nature of exposure is in turn determined by the environmental compartment, physical state of the toxicant and the nature of contact. The contact between the toxic compounds and organisms may be unintentional or accidental. In general, the chance of contact is the greatest in the immediate vicinity of pollution source and the pollutants enter the organism through a number of exposure routes.

Organisms are exposed to the toxicants through a number of routes such as food, air, skin, water and so on. Though the toxic effects are produced only after the absorption of toxicants, it is yet important to know the route by which the substance has entered into the body. The absorption of the substance is determined by the barriers present along the route of entry into the body. For example, the uptake of a substance via the lungs or intravenous injection causes the substance to enter the general circulation directly. But in the case of oral ingestion, the substance is first transported to the liver where it is converted into products that may produce more toxicity or may be less toxic. Table 4.1 shows the routes, sources and nature of toxicants.

Table 4.1   Routes, sources and nature of toxicants

Route	Sources	Nature	Entry point
Food	Natural — plants and animals	Direct	Mouth → gastro intestinal tract
	Artificial — contamination and addition	Direct or indirect	Mouth and through gills and skin
Water	Natural — minerals and organic compounds	Direct	Mouth → gastro intestinal tract
	Artificial — pollution and water discharge	Direct	Mouth → gastro intestinal tract
Air	Artificial — emission from various sources	Direct	Lungs
Skin	Intentional — medication	Administration	Skin
	Unintentional — polluted work place	Direct	Skin

ROUTES OF ENTRY OF XENOBIOTICS

Through Food

Many toxic chemicals are found to occur naturally in food in minute quantities so that they do not produce any adverse effects on animals. Though some edible plants contain highly toxic substances, they are degraded in the process of normal preparation. However, the foodstuff contain a number of undesirable substances such as pesticide residues, heavy metals, residues of medicines, etc. which are referred to as **contaminants**. The direct exposure of the contaminants is through consumption of crops grown in polluted soil and the indirect exposure is through eating of meat or milk from the animals that have been fed with polluted grass or feed. The excess use of nitrogen as a fertilizer causes accumulation of nitrate in vegetables and crop plants and this nitrate is converted into nitrite and nitrosamine, which are carcinogenic. The foodstuff are also intentionally added with additives which are harmful as they produce many adverse effects.

Through Water

Water forms an efficient route for the exposure of a number of potentially harmful chemicals. In aquatic animals, the toxicants enter into the body from the surrounding water along with food or through the skin or gills. A large quantity of water circulates through the gills so that the toxic substances enter the bloodstream very efficiently through the thin epithelial layer of gills.

Through Air

The indoor air is highly contaminated due to household activities whereas the outdoor air is polluted by industrial activities and combustion of fossil fuels. The toxic pollutants released into the air are easily dispersed over wide areas in a relatively shorter time.

Through Skin

In general, exposure of toxicants via skin is unintentional and is especially occupational. The penetration of certain lipophilic substances occurs through skin at high rates.

Through Egg Yolk and Placenta

The early developmental stages of animals are more susceptible to toxic substances. A number of lipophilic substances are shown to be present in the egg yolk of fishes and birds. In mammals, certain toxic chemicals or their metabolites are excreted in the milk so that the offsprings are affected indirectly through their mothers.

PENETRATION ROUTES IN LABORATORY ANIMALS

In experimental animals, the penetration of xenobiotics can be performed in the following ways.

Inhalation   By inhalation, the toxicants in the vapour or gaseous state enter into the organisms through respiratory surfaces.

Topical application   The toxicants in the liquid form are applied topically on the organism and are absorbed through the skin.

Injection method   The toxicants along with the carrier materials are injected into the body through intravenous, subcutaneous, intramuscular and intraperitoneal routes.

Dipping method   The organisms are simply dipped into the toxicant which is in the form of solution, suspension or emulsion.

Feeding method   The toxicant is mixed with the food and fed to the organisms (oral route) for a desired period.

CATEGORIES OF EXPOSURE BASED ON DURATION

The exposure of organisms to toxicants can be categorized into four groups. **Acute exposure** or **short-term exposure** involves the exposure of organisms to a chemical for relatively shorter duration, probably up to 96 hrs. The **subacute exposure** involves the repeated exposure of organisms to sublethal concentrations of the toxicants for about a few weeks to one month duration. The **chronic exposure** or **long-term exposure** involves the repeated exposure of organisms to sublethal concentrations of the toxicants for a longer duration.

The exposure routes for the above groups are usually intraperitoneal, intravenous, subcutaneous, oral or cutaneous. However, the most common route for chronic exposure is oral ingestion.

FACTORS AFFECTING EXPOSURE

The exposure of xenobiotics to the organisms depends on their physical and chemical forms, the medium in which they are present, their mobility in the soil, and the biology

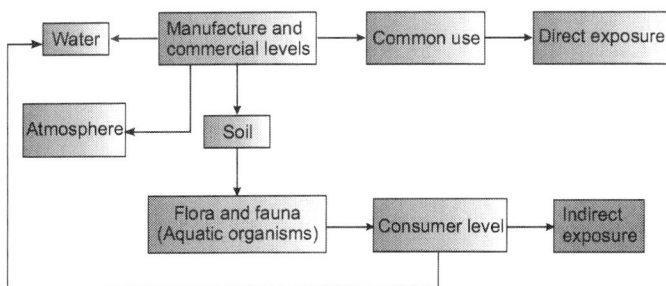

Figure 4.1   Routes of exposure of pesticides to man

of the organisms. Volatility also plays a significant role in the exposure of the toxicants into the organisms. The routes of exposure of pesticides to man are shown in Figure 4.1.

EXPOSURE RESPONSES AND MONITORING

The harmful effects of xenobiotics are determined by the environmental burden and the extent of exposure of organisms in that environment. In order to provide regulations concerning pollutants, it is important to know the quantitative relationship between pollutants and exposure of organisms in a particular environment. As the severity of the toxic impact exerted by toxic chemicals depends on the extent and duration of exposure as well as the sensitivity of the animal, the effects will vary widely, i.e., barely noticeable, reversible or irreversible disorders and lethal. In each of particular exposure, an effect will have certain intensity in a particular number of individuals. The percentage will be greater, if the extent and duration of exposure are longer. This is called **quantitative exposure response relationship**. The regular measurements of the level of exposure of animals and man to toxic substances over a period of time is known as **monitoring**. The extent of exposure can be determined by measuring the concentration of the substances in the environment and in the food. It can also be determined by measuring the level of the substance or its metabolites in one or more organs of the body as well as by determining the effects caused by the substances on organisms. The internal exposure can be measured by the determination of substances or metabolites in blood or urine and this is more suitable to the workers in industries. In general, exposure monitoring provides information on the intake of toxicants by the animals via all exposure routes together.

EXPOSURE RISK ANALYSIS AND PREVENTION

To make risk analysis for a substance, information on the occurrence of the substance in the environmental compartments (air, water, soil and biological compartments) and its behaviour within them are required. The government can formulate standards for chemicals based on toxicological as well as socio-economic interests. Before formulating a standard, assessment of exposure that takes place or might take place along with its potential consequences must be made. Then the present and future exposure must be analysed and compared with existing standards. Now the data can be used as the basis to carry out measures to be taken. A few standards have to be designed based on the effects that occur after exposure. The adverse effects of toxicants on animals and humans can be limited by reducing the exposure to these substances through various methods. The production and use of certain chemicals can be regulated, as in the case of pesticides which can be restricted either by withdrawing the licence or by banning the production. Certain pollutants can be degraded by using microbes, as practised in biological freshwater purification. The bioavailability of the substances can be reduced by physical and chemical means.

REVIEW QUESTIONS

1. Define the term "exposure" and enumerate the factors which affect the entry of toxic substances.
2. Explain the various routes of entry of xenobiotics.

3. Write notes on the following:
    i. Entry of toxicants through food
    ii. Entry of toxicants through water
    iii. Entry of toxicants through air
4. In what ways are the chemicals administered in experimental animals?
5. Categorize the exposure of organisms to toxicants.
6. Write an account on exposure monitoring and risk analysis.

5

ABSORPTION OF TOXICANTS

The absorption of toxic substances from the environment into the body of organisms mainly depends on their ability to cross semi-permeable membranes. In general, biological membranes are readily permeable to liphophilic or hydrophobic substances. The hydrophilic substances can also pass through the membranes but only at a slower rate. The water-filled pores of the cell membrane enable the hydrophilic substances to pass through the membrane. The biological barrier may also include larger complexes formed by closely associated cells as in the case of endothelial cells in which the intercellular junctions take part in the absorption of hydrophilic substances in addition to the pores of cell membrane.

MEMBRANE BARRIERS

A toxic substance produces harmful effects on organisms only when it is absorbed and transported to the site of action through circulation. Living organisms are protected from their environment by specialized membrane coverings which restrict the free access of toxicants into their body. Thus the toxicants have to pass through a number of membrane barriers to bring an impact in the body of organisms. The membrane barriers include a large number of cell types such as skin epithelium, membranes of lungs and gastrointestinal tract, capillary epithelium, etc. As an example, the blood–brain barrier is less permeable than other areas as the capillary endothelial cells of the central nervous system are tightly packed with much less protein content in the interstitial fluid and are surrounded by glial cell processes. In the placental barrier, there are many layers of cells between the foetus and maternal circulations. At tissue level, a toxic substance has to pass through the membranes of cell, mitochondria and nucleus. The membrane barriers found in living organisms are shown in Figure 5.1.

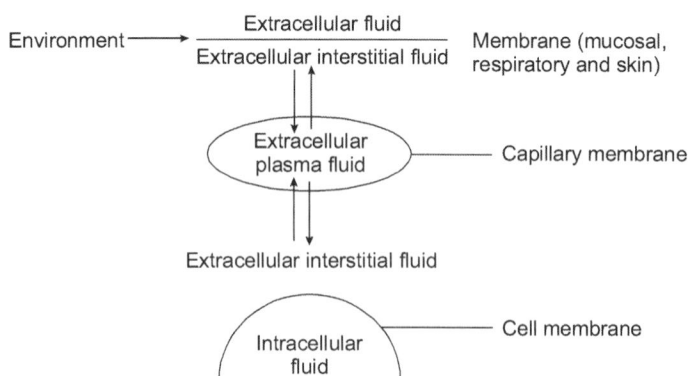

Figure 5.1   Membrane barriers found in living organisms

SITES OF ABSORPTION

Generally, the actual site of absorption depends on the mode of entry of toxicants. Toxic substances are usually absorbed by the organisms through three different means, namely, gastrointestinal tract, respiratory system and the skin.

For example, orally ingested toxicants are absorbed by the gastrointestinal tract; topically applied chemicals by the skin; volatile substances by the respiratory organs, and so on.

Absorption Through Digestive Tract

Absorption of xenobiotics occurs throughout the entire length of the digestive tract. Though mouth acts as a passage of entry for toxicants, it also absorbs certain toxic chemicals (**buccal absorption**). The wall of the gastrointestinal tract is the main route of absorption of many substances after the oral intake. The passage of toxic substances from the tract into the blood requires only the transport across the epithelial layer which provides a very large surface area of absorption due to the presence of villi and microvilli, especially in the small intestine. Therefore, the largest absorptive surface area is found in the small intestine through which most xenobiotics are absorbed. The absorptive surface area at different regions of the digestive tract are depicted in Table 5.1.

Table 5.1 Absorptive surface in the digestive tract

Regions	Absorptive surface area (%)
Mouth	0.02
Stomach	0.10–0.20
Small intestine	100.00
Large intestine	0.50–1.00
Rectum	0.04–0.07

The blood vessels from the gastrointestinal tract open into the liver via the portal vein so that the substances absorbed from the gut are first transported to the liver by the venous blood. In the liver, the toxic substances are mostly metabolized, removed from the blood and excreted into the bile or converted into their by-products and stored. Thus the portal vein system provides protection against poisonous substances. The toxic substance or one of its metabolites in the bile may be reabsorbed on entering into the intestine and is excreted once again into the bile. This process is known as **enterohepatic circulation** (Figure 5.2) which slows down the elimination of substances from the body.

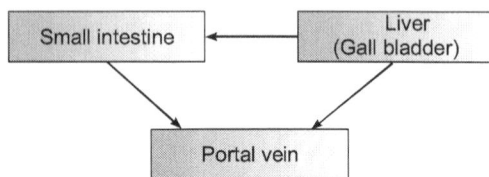

Figure 5.2   Schematic diagram of enterohepatic circulation

Lipophilic toxicants are more rapidly absorbed through the intestinal wall by passive diffusion. Some low-molecular weight substances pass across the intestinal membrane through the pores in an aqueous phase. The absorption of weak acids and bases depends on the pH of different regions of the intestinal tract. Hence the weak acids are absorbed by the stomach which possesses low pH and alkaloids are mainly absorbed by the intestine that is more alkaline. In general, highly dissociated and hydrophilic substances are not absorbed by passive diffusion through the intestinal wall and so a relatively small quantity of hydrophilic substances are absorbed by the digestive tract.

Absorption Through Respiratory Organs

In terrestrial animals, lungs form the important route for gaseous toxicants whereas the gills absorb non-volatile substances dissolved in water in aquatic organisms.

The lungs serve as an important route for the absorption of toxic gases and volatile solvents as well as aerosols (particles suspended in air). In lungs, absorption of substances occurs via the alveoli in which the epithelium is extremely very thin with a very large surface area for high diffusion of substances. In addition, the pulmonary capillaries also have a very large surface area with high perfusion efficiency. Thus the lungs with a thin air–blood barrier are the most effective absorptive areas in the body. The low-molecular weight lipophilic substances are rapidly absorbed by the lungs but the hydrophilic substances diffuse mainly through the water-filled pores. The extent of penetration into airways is determined by the molecular size of the suspended particles. In aquatic animals like fishes, the gills are the efficient organs to absorb environmental chemicals. The mechanism of absorption by the gills is very similar to the absorption of substances by the lungs. The gills not only provide a large surface area but also capable of absorbing non-volatile substances which are dissolved in water. In general, the gills absorb lipophilic substances from water more effectively.

Absorption Through Skin

Absorption of toxicants via the skin is called **percutaneous absorption** because the toxic chemicals have to reach the circulation from the outer surface of the skin through the outer nonvascular horny layers — the epidermis and dermis. In animals, the skin is a good and continuous lipid barrier, so it is relatively impermeable. Therefore, the toxicants have to pass across the keratinized skin layer to reach the bloodstream. It is shown that the skin can absorb lipophilic compounds of low-molecular weight very quickly and also some small hydrophilic substances. The percutaneous absorption of toxic chemicals occurs primarily at work places where the people confront with a wide variety of toxicants. This depends mainly on the physical and chemical properties, degree of ionization, molecular size and solubility of the substances. Local factors such as pH, temperature, etc. can also influence the absorption of chemicals through the skin.

Absorption routes in experimental animals When the toxicants are administered into the experimental animals in laboratories through different injection methods, they can be absorbed through intraperitoneal, subcutaneous, intramuscular and intravenous routes.

MECHANISM OF ABSORPTION OF TOXICANTS

The toxicants pass through the membranes of organisms by several methods as shown in Table 5.2.

Table 5.2   Transport mechanisms of toxicants

Mechanism	Nature of substances
Diffusion through lipid membrane	Hydrophobic substances
Diffusion through pores	Small hydrophilic substances
Filtration	Small hydrophilic substances
Facilitated diffusion	Substances which cannot bind to carriers
Active transport	Substances which cannot bind to carriers
Phagocytosis and pinocytosis	Larger molecules

Simple diffusion It depends on the concentration gradient without requiring energy so that there is no resistance for the absorption of materials in membrane systems. The small hydrophilic substances diffuse through lipid membranes in aqueous regions and the large organic substances through lipid regions. The lipophilic substances and lipid-soluble unionized substances can easily diffuse across the cell membranes but the ionized forms with low lipid solubility are not able to diffuse into the cell membranes.

Filtration The substances having low-molecular weight which are readily soluble in water and penetrate through the membrane pores along with water molecules.

Active transport It occurs against concentration or electrochemical gradient and requires energy. By this process, the toxicants combine with carrier macromolecules to form a complex substance which penetrates across the cell membrane.

Facilitated diffusion This process is also carrier-mediated but does not occur against the concentration gradient.

Phagocytosis and pinocytosis By these two mechanisms, the cell membrane engulfs the toxicants.

KINETICS OF ABSORPTION

Rate of absorption The rate of absorption of toxic substances into the organisms mainly depends on the nature of chemicals and their site of administration. The unionized lipid-soluble molecules of substances readily penetrate the cell wall whereas the rate of absorption is slower for strong acids and bases. Similarly, the absorption of toxicants through the skin is very slow, through the lungs is rapid and through the gut is complex.

In general, absorption of chemicals occurs either as a **first-order process** (at low doses, the rate of reaction is directly proportional to the amount of toxicant present) or a **zero-order process** (as the concentration of the substance increases, a point may be reached at which there is no further increase in the rate of absorption). In some instances, the rate of absorption does not follow the above two processes. For example, the lipophilic substances are absorbed rapidly but eliminated slowly. The highly polar molecules are absorbed by the gut slowly but are eliminated rapidly from the blood via the kidneys.

Extent of absorption   This depends on the bioavailability of a substance which is a fraction of the dose absorbed into the circulation from the site of exposure.

REVIEW QUESTIONS

1. How do the membrane barriers play a role in bringing about a toxic impact?

2. Explain the three major routes of absorption of toxicants.

3. Brief the following.
   i. Absorption through digestive tract
   ii. Absorption through respiratory organs
   iii. Absorption through skin

4. Give an account of the mechanisms involved in the absorption of toxic chemicals.

5. What are the factors which influence the rate and extent of absorption of toxic substances?

6

DISTRIBUTION OF TOXICANTS

Consequently upon absorption, the toxic substance enters into the blood or intravascular space, then into the intercellular space and membrane of the tissue cell, and finally into the membrane of the cellular components. The intravascular space is separated from the intercellular space by the cells of vascular wall. In organisms the accessibility of extravascular space to the toxic substances varies from one region to another because the structure and composition of this wall is not uniform throughout the organism. Generally, the distribution of xenobiotics from blood circulation to various organs depends on the blood flow into the organs.

After the entry into the circulatory system, the toxic substances are distributed throughout the body via the bloodstream. The absorbed substances either accumulated near the site of action or are stored for sometime and transported to the organs. Initially, the toxicants are distributed to various aqueous compartments by simple diffusion. Some are exclusively distributed over the blood and interstitial fluid so that they fill only the extracellular fluid compartments without crossing cell membrane.

Many small hydrophilic molecules, which are insoluble in lipids and are able to cross the cell membrane, are homogeneously distributed over both extracellular and intracellular compartments. Some substances are distributed over both aqueous and lipid compartments (mixed distribution). Many substances are found in much higher concentrations in tissues than in blood plasma due to binding to cellular components or dissolution in lipids (accumulation or sequestration). For example, highly lipophilic substances that cannot be metabolized selectively accumulate in the body fat. In general, most of the lipid-soluble substances are homogeneously distributed over various organs whereas the concentrations of more polar substances are much higher in organs like liver and kidney. These organs have high-accumulative capacity so that they are subjected to toxicological examination in intoxication cases. The membranes and fluid compartments involved in the distribution of xenobiotics are shown in Figure 6.1.

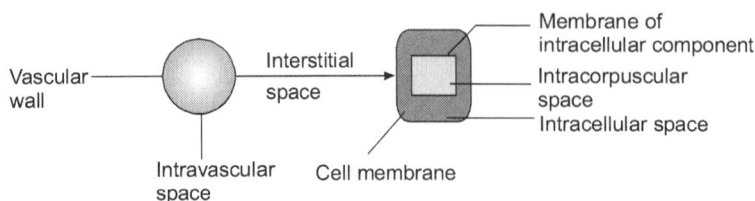

Figure 6.1 Schematic representation of membranes and fluid compartments in distribution of toxicants

ROLE OF PLASMA PROTEIN BINDING IN DISTRIBUTION

In general, many xenobiotics dissolve partly in the plasma and are distributed throughout the organism in partially dissolved condition. But an important distribution mechanism for toxic substances is their ability to bind with plasma proteins. This binding may either be reversible or irreversible. Most of the toxic chemicals or their metabolites bind reversibly to a variety of biological components. Many substances bind with plasma proteins strongly, leaving only a small fraction in the plasma in free unbound form. In general, organic anions have more strong binding capacity than organic cations. The binding of xenobiotics to plasma proteins reduces the concentration of the substances in the blood, thus decreasing the intensity of the toxic effect and protecting the tissues. At the same time, this binding reduces or slows down the elimination of substances. As a result, the blood proteins provide a kind of reservoir for the toxicants so that the duration of the effect is prolonged. The reversible nature of binding would cause competition for the binding sites. This results in the displacement of biological molecules by the toxicants causing harmful effects in organisms. The binding capacity of plasma proteins with toxic substances varies in different species and this accounts for the differences in the toxicity found between species.

PLASMA CONCENTRATION OF TOXICANTS

As the blood forms the transport medium of toxic substances in the body of organisms, the concentration of substances in plasma concentration is very important because it indicates the distribution of toxicants to various tissues. For example, those substances which are readily soluble in lipids and distributed to various tissues of the body are less concentrated in the plasma. On the other hand, the substances which are readily distributed in the tissues are found in higher concentrations in the plasma.

STORAGE DEPOTS

During distribution, the toxicants are concentrated or accumulated in a particular tissue, organ or any other site. Such areas are called storage depots. They offer protective mechanisms to the organisms by preventing the accumulation of substances at their site of action, thus reducing the toxicity of chemicals. If the toxicants are highly concentrated at their sites of action, then their impact is highly toxic, whereas only less effects are evident if they are concentrated elsewhere. In storage depots, the toxicants interact with the plasma proteins so that the binding could affect the distribution. This is because the macromolecules, to which the toxic substances are attached, restrict the entry across the membranes.

Kinds of Storage Depots

Plasma proteins   The toxic substances are capable of binding reversibly with the plasma proteins, especially albumin through hydrogen bonds, van der Waal's forces and ionic bonds. As the proteins are high-molecular weight substances, they cannot pass through the capillary walls and cell membranes. As a result, the substances which are bound to plasma proteins are restricted from reaching their site of action. Thus binding of toxicants with protein molecules of the body constitutes **storage depot** located throughout the body of organisms. A reduction in the concentration of free form of chemicals in the plasma will cause their release from the storage depot so that there is an equilibrium between free and bound forms of toxicants. But the toxicity of a substance is increased when the substance bound to the plasma protein is displaced by another toxicant.

Fat   The body fat acts as a storage depot for the highly lipophilic substances which are distributed and accumulated easily in it. Consequently, the concentration of toxicants is lowered in target organs.

Liver and kidney These are the two major organs which accumulate large quantities of toxic substances with the help of intracellular-binding proteins. In these organs, the chemicals are biocatalytically transformed into hydrophilic substances which are easily excreted.

Bone Some toxic substances are also found to accumulate in bones which serve as main storage sites.

KINETICS OF DISTRIBUTION

The rate of distribution is rapid in the plasma when a toxicant is administered through a single intravenous dose but is slow in the tissues. The rate of distribution of a substance from the blood into the body tissues can be quantified only when it enters the circulation at a greater rate in comparison to the rate of its distribution. In general, the distribution of toxic substances in the body of organisms is mainly instantaneous.

The extent of distribution is related to the apparent volume of distribution (*V*). It represents the total quantity of toxicants in the body and plasma.

$$V = \frac{Ab}{C}$$

where,

C refers to the plasma concentration and

Ab refers to the total body load.

The volume of distribution represents the actual space in which a toxicant is distributed.

$$V_d = \frac{D}{C_p}$$

where,

V_d refers to the volume of distribution,

D refers to the dose administered and

C_p refers to plasma concentration.

REVIEW QUESTIONS

1. How does plasma concentration determine the distribution of toxicants?
2. Write an essay on storage depots.
3. What are the factors which influence the rate and extent of distribution of toxicants?
4. Explain the role of plasma proteins in the distribution of toxic substances.

7

EXCRETION OF TOXICANTS

When a toxic substance is absorbed and distributed throughout the body of organisms, it is usually removed from the body. This process is called **elimination** which decreases the quantity of the toxicants in the organisms and also reduces the amount of the substance at its site of action. The foreign substances are totally removed from the body via two major processes, namely, excretion and biotransformation. **Excretion** is the elimination of a foreign substance or its metabolites from the body of organisms. The mechanism of excretion is mainly responsible for most of the elimination processes. Therefore, excretion is an important step by which the toxicants are eliminated from inside the body of organisms and are inhibited from producing adverse effects. The most important excretion routes include **renal excretion** via the kidneys into the urine and **biliary excretion** via the bile into the faeces. The volatile substances are mainly excreted through the lungs. The direct excretion via faeces, and excretion through saliva, sweat, tears, nasal mucus and milk are lesser important routes for the elimination of xenobiotics. Generally, the hydrophilic substances are excreted as such whereas the lipophilic substances are removed only after they have been converted into hydrophilic metabolites.

RENAL EXCRETION OR URINARY EXCRETION

It is one of the most important routes by which the toxic chemicals and their metabolites are excreted via three processes, namely, glomerular filtration, tubular secretion and tubular reabsorption as shown in Figure 7.1.

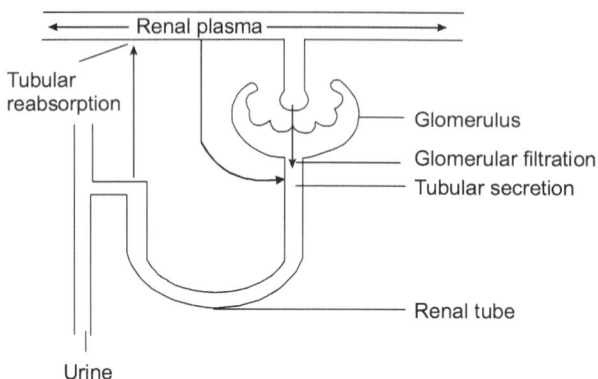

Figure 7. 1   Diagram showing the process of renal excretion

Glomerular Filtration

As kidneys receive nearly 25% of blood from the cardiac output, they filter a significant quantity of toxicants. It involves ultrafiltration by which substances having molecular weight ranging between 5,000 Da (Daltons) (man) and even 40,000 Da to 60,000 Da (rats) are filtered through the glomerular membrane. However, a high-molecular weight complex formed by binding of the toxicants with plasma proteins cannot be filtered through glomerulus. The lipophilic substances diffuse into the tubules from the blood by active transport but are reabsorbed into the blood by passive diffusion. But small hydrophilic substances are excreted via urine from the blood through filtration (glomerular filtrate) so that they are not reabsorbed by the tubules and are easily eliminated out of the body through urine.

Tubular Secretion

This involves various highly specific active transport systems in the proximal tubes. In addition, two other separate systems, namely, **anionic system** and **cationic system** with less specificity are responsible for tubular secretion. The anionic system transports organic acids, sulphonic acids and acidic metabolites whereas the cationic system transports organic bases and quaternary ammonium compounds. As these systems possess low specificity, they are capable of actively transporting a number of different compounds into the tubule. Moreover, the substances, which bind to plasma proteins, can also be transported into the tubules.

Tubular Reabsorption

Tubular reabsorption is mediated by specific transport mechanisms, which transport endogenous substances like glucose and amino acids. Therefore, the xenobiotics cannot be actively reabsorbed unless they are closely related to the specific endogenous compounds. That is why, the toxicants are mostly reabsorbed by passive diffusion through a concentration gradient. As the ultrafiltrate concentrates at high level in the renal tubule, a large proportion of these substances, especially lipophilic ones, diffuse back into the blood. As a result, the non-volatile and lipophilic substances can be excreted only from the body with great difficulty unless these compounds are converted into more polar metabolites by biotransformation. The reabsorption of organic acids and bases in the tubule mainly depends on the ionization ability of the substance and the pH of the environment.

Thus the net renal excretion depends on the physico-chemical properties of the toxic substances and on the functioning of the transport mechanism in the renal system.

BILIARY EXCRETION

It is a complementary route to renal excretion of xenobiotics. In other words, smaller molecules are eliminated via the kidneys, whereas larger molecules through the bile. Thus the biliary excretion forms an important route to eliminate many conjugated compounds especially the polar substances. The transfer of toxic substances from blood to bile is an active process involving three separate systems, namely, one for acids, one for bases and one for polar substances which cannot dissociate. The biliary excretion increases with molecular weight of the substances whereas the renal excretion correspondingly decreases. The influence of molecular weight of biphenyls on their route of excretion is shown in Table 7.1.

Table 7.1   Molecular weight of various biphenyls and their route of excretion in rats

Molecular weight	% Excretion	
	Biliary	Urinary
154	80	20
188	50	50
223	34	66
326	11	89
361	1	99

The dependency of biliary excretion on molecular weight is in turn species-specific. Other compounds which are excreted via the bile can also inhibit the excretion of toxic substances. After reaching the intestine, the substances are excreted as unchanged molecules through faeces, or reabsorbed as unchanged molecules or undergo enzymatic conversion in the intestine followed by reabsorption into the blood. The products of enzymatic reaction in the intestine may be excreted directly via the faeces or reabsorbed. All these events are collectively called as **enterohepatic circulation.** This process certainly has toxicological impacts when there is reabsorption of metabolites which are more toxic than their parent compounds. It may also increase the half-life of xenobiotics in the body.

EXCRETION VIA LUNGS

The elimination of toxic substances through the lungs requires no special transport mechanism and occurs simply by diffusion. The substances in the gaseous or volatile state are mainly excreted by the lungs. The concentration of the substance in plasma and alveolar air and the blood–gas partition coefficient are the major factors which influence the excretion of substances via the lungs. The rate of excretion of gases is inversely proportional to the rate at which they are absorbed. In general, the gases which dissolve easily in the blood are excreted very slowly by the lungs whereas the gases which do not dissolve in blood are excreted rapidly. Therefore, it is an efficient route for the excretion of volatile liphophilic substances.

EXCRETION VIA FAECES

In a true sense, excretion of toxic substances via the faeces is rather due to biliary excretion. Direct excretion through faeces occurs only when the substances are incompletely absorbed or not absorbed at all. The substances which are secreted through saliva and digestive juices as well as the inhaled substances which are swallowed from the respiratory tract are also excreted through the faeces. It is found that the direct excretion from blood through the intestine by passive diffusion forms an important route for the lipophilic compounds to be excreted.

EXCRETION VIA SWEAT AND SALIVA

The non-ionized and lipid-soluble substances are generally excreted by simple diffusion. The substances which are excreted through the saliva enter the alimentary canal and are eliminated through the faces.

EXCRETION VIA MILK

It has been shown that many toxic compounds are excreted through milk which is reported to contain higher concentrations of the substances than found in the maternal plasma. The non-ionized substances are transported to the mother's milk by passive diffusion. It mainly depends on the concentration of the substances, pH, the plasma concentration and lipid solubility.

EXCRETION VIA EGGS

The excretion of toxic xenobiotics does not affect the mother animal but greatly affects the survival of the young ones. In all oviparous animals, the toxic substances are mainly excreted via the eggs which accumulate the polar compounds in the egg white and the liphophilic xenobiotics in the yolk. Generally, these animals do not convert the toxicants into less toxic substances. The factors which influence the elimination of toxicants via milk also play a role in the elimination via the eggs.

EXCRETION VIA PLACENTA

The toxic substances which have low molecular weight and lipophilic substances of medium molecular weight will pass across the placenta by simple diffusion. The differences in placental structure do not influence the transport of substances so that the concentration of toxicants in the foetal tissues are the same as in the maternal tissues. Experiments have proved that the lipophilic substances accumulate in foetal liver, adipose tissue and alimentary canal.

EXCRETION VIA OTHER ROUTES

The excretion of toxicants via the hair, nails and skin is not of much importance. However, some toxic metals are found to be stored in these tissues.

KINETICS OF EXCRETION

The rate of elimination of a substance from the body indicates the rate of change of its body load. The rate of excretion mainly depends on the ability of the organs to extract the chemicals from circulation and to remove it either by metabolism or excretion and the extent to which the chemicals remain in circulation to be available for elimination. In short, the rate of elimination is determined by two factors, namely, the apparent volume of distribution (V) and the clearance (CL).

Plasma Clearance

Clearance is defined as the ratio of rate of elimination of the chemical to the plasma concentration. This is given by

$$CL = \frac{\text{Rate of elimination of the chemical } (\mu g \text{ min}^{-1})}{\text{Plasma concentration } (\mu g \text{ ml}^{-1})}$$

Plasma clearance represents the sum of all individual clearance processes such as metabolism (CL_M), renal excretion (CL_R), biliary excretion (CL_B), etc.

$$CL_P = CL_M + CL_R + CL_B ...$$

Metabolic and renal excretion are the major routes of total clearance. The renal clearance is the ratio of rate of elimination of a substance in the urine (CL_R) to the concentration of the substance in the plasma and is represented as

$$CL_R = \frac{\text{Rate of elimination in urine} (\mu g \, min^{-1})}{\text{Concentration of the substance in plasma} (\mu g \, ml^{-1})}$$

The metabolic clearance cannot be measured easily, so it can be derived from total clearance and renal clearance by the following formula.

$$\text{Total clearance} (CL_T) = CL_M + CL_R$$
$$CL_M = CL_T - CL_R$$

REVIEW QUESTIONS

1. Give a brief account of the most important elimination routes of xenobiotics.

2. Explain renal or urinary excretion of xenobiotics.

3. Explain biliary excretion of xenobiotics.

8

BIOTRANSFORMATION

As far as toxicology is concerned, the term metabolism can be used as a synonym for biotransformation because both processes involve the biological conversion of substances. While, metabolism is concerned with useful substances, biotransformation involves xenobiotics. The metabolic transformation of toxicants is known as detoxication because they are converted into less toxic products. The organisms also convert the toxicants into more toxic substances and the metabolism of such compounds is called as activation. The conversion of one toxic compound in the body of an organism into another poisonous substance through biological catalysis is biotransformation; the parent compound is transformed into more metabolites through a series of enzymatic reactions. These products possess distinct physico-chemical properties and hence they have entirely different toxic characteristics from their original compound.

In organisms, the polar or hydrophilic compounds are eliminated quickly. The non-polar or lipophilic compounds are gradually absorbed and accumulated as they are not easily eliminated. Therefore, the lipophilic substances are more often altered into the hydrophilic forms biocatalytically so as to enable easy excretion. Through biotransformation reactions, an active chemical can be converted into another inactive or active substance and vice versa. The reactions of biotransfomation are mostly catalysed by the enzymes found in many organs of the body, for example, the liver plays an important role in the biotransformation of xenobiotics.

BIOTRANSFORMATION SYSTEM

In an environment, the organisms are exposed to a large number of chemical pollutants which cause adverse effects on them. In general, the living organisms eliminate the xenobiotics from their body either by direct excretion or by metabolic transformation of the toxicants. The metabolic pathways responsible for the transformation of biologically active substances into their metabolites are called **biotransformation system.** In other words, organisms possess a number of enzyme systems capable of converting the foreign substances and endogenous waste products into more water-soluble substances, which are more readily excreted. The processes involved in these conversions constitute the biotransformation system. This system includes phase I reactions (oxidation, reduction and hydrolysis) which introduce a polar group into the molecule and phase II reactions (conjugation) which conjugate a hydrophilic substance with a polar group in a molecule, resulting in the formation of water-soluble conjugated products. A general overview of biotransformation of xenobiotics is given in Figure 8.1.

Phase I Reactions

In phase I reactions, the functional groups of toxic substances (OH, SH, NH_2, COOH, etc.) are either added or exposed.

They include oxidative, reductive, hydrolytic and other enzymatic reactions in order to convert the non-polar compounds into polar end products.

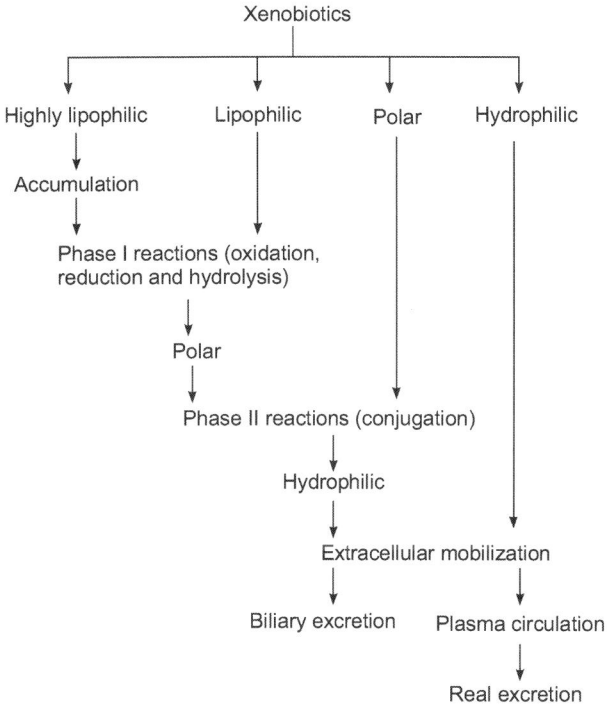

Figure 8.1 An overview of biotransformation of xenobiotics

Oxidative reactions The oxidative reactions, also collectively known as mixed function oxidase (MFO) system, include many important reactions catalysed by enzyme systems such as cytochrome P-450. These enzymes are mostly found in microsomes of various tissues especially in the liver. The reactions require NADPH as a cofactor, involve an electron transport system and oxidize many different substrates. During oxidation, the toxic substrate combines with the oxidized cytochrome P-450 which gets reduced by getting an electron from the NADPH–cytochrome-*c*-reductase. The reduced P-450 now reacts with CO to form one or more complexes. Though the mechanism of the above reactions are not clearly understood, the following steps can be recognized in the oxidation of the substrate by cytochrome P-450 system.

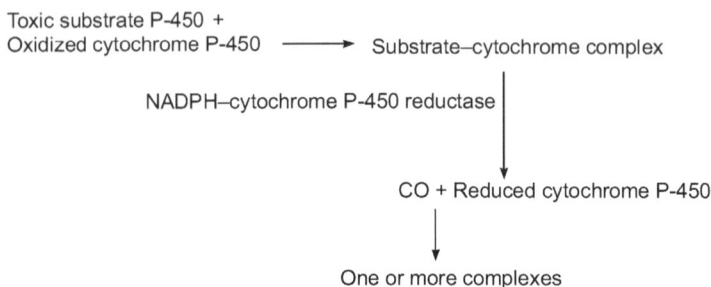

Toxic substrate P-450 +
Oxidized cytochrome P-450 ⟶ Substrate–cytochrome complex

NADPH–cytochrome P-450 reductase

CO + Reduced cytochrome P-450

One or more complexes

Here, each system with cytochrome P-450 is called mono-oxygenase. As a single system contains many cytochrome P-450, each can react with different substrates. Therefore, the entire system can oxidize a number of xenobiotics. Thus many mono-oxygenases together constitute MFO system. If the hydrophobic or lipophilic substrate binds with a lipophilic site of cytochrome P-450, then it forms **Type I binding**. In the case of organic nitrogen-containing substrate, haem-ion binding occurs which is called **Type II binding**. This MFO system catalyses deamination, demethylation, dealkylation, aromatic ring hydroxylation, alkyl and *N*-hydroxylation, cleavage of ester bonds, epoxidation and oxidation of sulphides, phosphorothionates, alcohols and aldehydes.

Reductive reactions   These are general biochemical reactions for certain xenobiotics but are not as common as oxidative reactions. In general, the reductive reactions may either be NADPH-dependent or NADPH-independent. For example, the aromatic nitrogroups are converted into corresponding amines by NADPH-dependent nitro reductase reaction. On the other hand, dechlorination of DDT into DDE occurs through NADPH-independent reactions which may occur either in the presence of enzymatic systems or in their absence. Thus the difference between these two types of reduction reactions cannot be distinguished. Therefore, the reduction reactions may be categorized into **enzymatic reduction reactions** involving various enzyme systems and **non-enzymatic reduction reactions** involving both chemical and biochemical interactions. The chemical reactions require biologically derived substances like vitamin B_{12} and CO II, whereas the biochemical reactions depend on a number of haemoproteins.

In higher animals, the reduction reactions include reductive dehalogenation in which a halogen atom is replaced by a hydrogen atom; reduction of nitrogroups to amines; conversion of pentavalent arsenics to trivalent arsenics, sulphoxides to sulphides, and aldehydes to secondary alcohols; *N*-demethylation; *N*-oxidoreduction and saturation of double bonds.

Hydrolytic reactions (esterases) The hydrolases which convert the esters into alcohols and acids in the presence of water are called **esterases**. These are not stimulated by cofactors but require divalent cations. Two types of esterases are important with reference to the metabolism of insecticides, namely **carboxylesterases** and **phosphatases**. The carboxylesterases metabolize the insecticidal esters through hydrolysis and the phosphatases metabolize the insecticidal esters through phosphorylation and dephosphorylation.

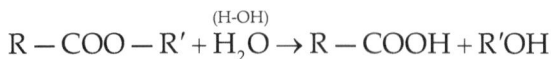

$$R - COO - R' + \overset{(H-OH)}{H_2O} \rightarrow R - COOH + R'OH$$

Phase II Reactions

These reactions are limited in number when compared to other metabolic reactions of xenobiotics and depend on some chemical centres and the biochemical capability of the animal. This is because the part of the molecule synthesized (conjugation agent) is provided by the animal itself. That is why some of these reactions are found only in some groups of animals and others occur in almost all species. The conjugation system mostly converts the apolar hydrophobic substances into water-soluble compounds which can be easily excreted through the routes of bile–faeces or kidney–urine systems.

Glucuronide or glucuronic acid formation   The formation of glucuronides is the most important phenomenon in phase II reactions, which occurs in almost all animals except insects. The polar group of substrate can conjugate only with activated glucuronic acid which consists of uridine diphosphate glucuronic acid (UDPGA).

UDPGA serves as a substrate for the enzyme **glucuronil transferase (GT)** which catalyses the conjugation process of a polar substance with glucuronic acid.

Phenol + UDPGA ⟶ Phenyl glucuronide

Similarly, compounds containing hydroxyl, carboxyl, amino, sulphydryl groups, etc. can also form glucuronides. The enzymes responsible for glucuronide formation are found in the liver of animals.

Glucose1-phosphate

+

UTP

Pyrophosphorylase

UDP-glucose

UDP-glucose dehydrogenase

UDP-glucuronic acid

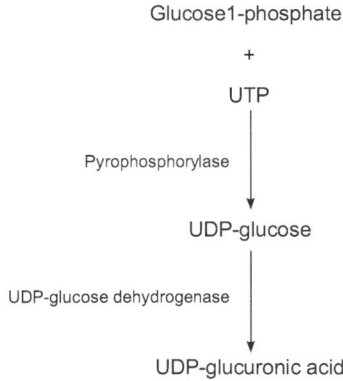

Sulphate conjugation This is another phase II reaction in which the sulphate is first converted to adenosine 5′-phosphosulphate (APS) which is then converted into 3′-phosphoadenosine-5′-phosphosulphate (PAPS).

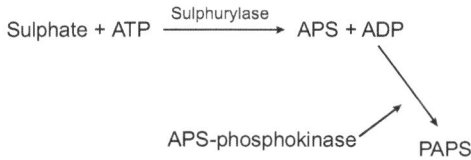

Sulphate + ATP $\xrightarrow{\text{Sulphurylase}}$ APS + ADP

APS-phosphokinase   PAPS

Now the substrate, for example phenol, combines with this activated sulphate (PAPS) catalysed by sulphotransferase.

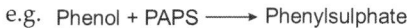

Substrate (phenol) + PAPS ⟶ PAP Sulphate

e.g. Phenol + PAPS ⟶ Phenylsulphate

Glutathion conjugation A number of substances can also be activated by glutathion conjugation. In some glutathion-mediated systems, glutathion is used as a catalyst in which the substrate directly binds with it where the glutathion level does not change at the end of the reaction.

DDT + Glutathion ⟶ Intermediate substance

↓

DDE + Glutathion

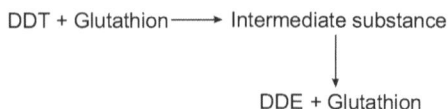

In some other systems, glutathion directly binds with the substrate with the help of glutathion-S-transferases. For example, in mammals, the activation of glutathion S-transferases require glutathion. In these systems, the conjugated product is converted into mercapturic acid through reactive intermediates with glutathion.

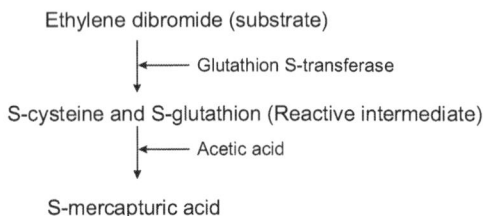

Ethylene dibromide (substrate)

|⟵ Glutathion S-transferase
↓

S-cysteine and S-glutathion (Reactive intermediate)

|⟵ Acetic acid
↓

S-mercapturic acid

The mercapturic acids are more excretable compounds.

Other conjugation processes **Acetylation** and **methylation** are the other two processes in phase II reactions in which the substrates like amines are metabolized. Acetylation is a general reaction involving aromatic amines and unnatural aromatic acids. Methylation constitutes normal metabolic processes of natural, primary, secondary and tertiary amines.

Through the phase II reactions, highly water-soluble products are produced and are easily eliminated from the body of organisms. Moreover, the phase II reactions are **bioinactivation reactions** or **detoxication reactions**. Thus biotransformation causes alterations in the biological activity of a xenobiotic compound. If this biological activity is decreased, it will lead to bioinactivation or detoxication. If this activity is increased then it will cause bioactivation.

DETERMINING FACTORS OF BIOTRANSFORMATION

The process of biotransformation plays a key role in the elimination of xenobiotics from the body so that the organisms are protected from the ill effects of toxicants. The metabolic reactions involving xenobiotics are complicated and are influenced by a number of factors. As the toxic impact would be similar both qualitatively and quantitatively in test animals and humans, it is very important to know the ways by which the species differ from each other in order to extrapolate the data from one species to another.

The following account gives the major factors which influence the biotransformation of toxic substances.

Species The biotransformation reactions vary from one species to another and even between the individuals of the same species. The variations may either be **interindividual** (between the individual of one species) or **intraindividual** (within one individual).

Genetic influences It is shown that a single gene regulates the oxidative metabolism of a large number of compounds. In humans, a bimodal population namely **extensive metabolizers** and **poor metabolizers** has been recognized based on the hydroxylation process. The same is also found to be true for phase II reactions based on which the human populations are categorized into **fast acetylators** and **slow acetylators.**

Sex The influence of biotransformation reactions by the sex of the organisms seems to be a type of genetic control but occurs via hormones. This, in turn, depends on the enzyme concentrations and on the lipid content of the enzymes.

Age   The phase I and phase II reactions are found to increase linearly from birth to adult stage due to cellular and molecular changes.

External factors   While most of the interindividual differences with regard to biotransformation reactions depend on species, genetic factors, sex, age, and so on, the intraindividual variations mainly depend on external factors and diseases. For example, certain diseases would affect the metabolic transformations by altering the physiology of the organs (liver, intestine, lungs and kidneys) which are involved in biotransformation process.

BIOINACTIVATION AND BIOACTIVATION

Generally, the toxic substances are made into water-soluble products by biotransformation reactions for being readily excreted. As a result, the substances do not bring out any toxic effects because the original toxicants disappear from the body of organisms due to the excretion of their metabolites. This process is called **bioinactivation** or **detoxication**. However, many phase I reactions yield products having higher toxicity than their parent compounds. When the substance adversely affects the components of biological systems such as proteins, DNA, etc., then it produces toxic impact, if present in higher concentrations. This process of formation of products that possess higher toxicity than the original substances is called **bioactivation**.

The production of hydrophilic substances by phase I reactions could not apply for most substances and so further reactions are required. In phase II reactions, the toxic substances are made into hydrophilic substances which are readily excreted. The introduction of polar groups during phase I reactions may undergo conjugation in phase II reactions; the products of which are highly water-soluble so that they are rapidly excreted. Therefore, phase II reactions are mostly bioinactivation reactions. As bioactivation and detoxication processes are metabolic reactions, the factors that determine biotransformation can also influence the above reactions.

REVIEW QUESTIONS

1. Define "detoxication" and "activation" of toxicants.
2. Write an essay on metabolism of xenobiotics.
3. Explain the reactions involved in primary metabolic processes of toxic substances.
4. Describe the MFO system.
5. Give a brief account of reduction and hydrolytic reactions in non-synthetic processes of biotransformation.
6. Explain different types of conjugation phenomena in the biotransformation process.
7. Give an account of biotransformation system.
8. What are the factors that determine the biotransformation reactions?
9. Define and explain the terms "bioactivation" and "bioinactivation".

9

MODE OF ACTION OF XENOBIOTICS

The action of toxicants in bringing overall changes in physical and chemical processes of organisms leading to mortality is known as mode of action. In other words, the impact of a toxic substance on the vital systems of organisms causing death refers to the "mode of action" of that toxicant. The action of xenobiotics may bring external or internal damage to any part of the organism. By entering into the circulatory system, the toxicants cause impairment of its function. They may also be taken by the blood and distributed to various tissues or organs whose functions are also impaired. Thus the toxic substances may either produce direct, intermediate or remote action on the target organisms. Some kinds of toxicants may act in more than one of the above methods. Moreover, the biological systems are so complex that it is difficult to elucidate the mode of action of toxicants and to classify them.

CLASSIFICATION OF TOXICANTS BASED ON MODE OF ACTION

Though the classification of mode of action of toxicants is difficult, it is possible to classify various toxic substances based on their mode of action on organisms.

❖ **Physical poisons**   These are the toxicants which cause mortality of the animals through physical effects, e.g. oxygen depletion leading to asphyxation.

❖ **Protoplasmic poisons**   These cause destruction of cellular protoplasm by the way of precipitating proteins, e.g. heavy metals.

❖ **Respiratory poisons**   These inhibit the catalytic action of respiratory enzymes thus reducing or blocking the cellular respiration in animals, e.g. carbon monoxide.

❖ **Nerve poisons**   These include lipid-soluble toxicants and actively inhibit the activity of acetylcholine esterase in animals, e.g. organochlorine, organophosphate and carbamate insecticides.

❖ **Haemotoxic poisons**   These reduce the production of blood cells from the haemopoietic organs and inhibit the functions of blood cell types, e.g. tannery, paper and pulp mill effluents.

THEORIES ON MODE OF ACTION OF TOXICANTS

The following theories have been put forward to explain the mode of action of biologically active chemicals.

The physical hypothesis   According to this theory, the physical disruption of membrane permeability is the cause for the toxicity of a substance. It gives much importance to the penetration and the concentration of a toxicant at the site of activity. However, this factor alone cannot be accounted as the main feature for the ultimate toxicity of toxicants.

The toxophore hypothesis   According to this theory, the toxic effect of a substance is due to a close link between toxophoric group of organisms and the chemical structure of the toxicant molecule.

The substitution hypothesis   According to this theory, some chemical substances which have similar properties to that of toxicants may cause disruption of metabolic pathway in animals.

The enzyme inhibition hypothesis   According to this theory, the toxicity of a substance is mainly due to its action on enzyme or enzyme systems of organisms. This is the situation found in almost all xenobiotics.

Compartment Model and Mode of Action

The manifestation of a toxic effect in the body of organisms may either be due to a reversible reaction between a toxic chemical and a target molecule or due to an irreversible interaction between them. In the first case, the intensity of the impact depends on the extent of binding to the target molecule or on the concentration of the toxic substance. The concentration of the substance at the site of action is determined by the blood concentration which in turn depends on its absorption, distribution, biotransformation and excretion.

The body of organisms can be assumed to consist of one or more compartments and to represent compartment model with all its tissues, organs and fluids that do not differ in terms of kinetics. One compartment model is the simplest one in which the body is considered to be a single homogeneous unit having relatively rapid distribution of substances with respect to absorption and elimination. However, the body behaves as consisting of at least two compartments when the exogenous substances are intravenously injected. In such cases, the substances are found to spread instantaneously in one compartment but are distributed over the second compartment more slowly. Here, the body consists of **central** and **peripheral compartments**, each having its own apparent volume of distribution. These compartments can be designated as **plasma compartment** and **tissues compartment** respectively. In many cases, the central compartment includes the circulatory system along with rapidly perfused tissues like liver, heart, lungs, kidneys and endocrine organs. The peripheral compartment constitutes the body fat along with poorly perfused tissues like skin and muscles.

Figure 9.1 explains the two compartment model of an organism and the fate of the substance that has been intravenously administered. When a substance is intravenously administered into the central compartment, it is rapidly spread over it. At the same time, the substance begins to disperse in the second compartment (peripheral compartment) with the simultaneous elimination of a proportion of the substance from the central compartment. In the beginning, there is a rapid net loss of the substance from the central compartment and a rapid decline in plasma concentration. This is due to the irreversible elimination from the central compartment through biotransformation, excretion and the uptake of the substance by the tissues of the second compartment. Then the substance is steadily distributed over various tissues and fluids, depending on the rate of transport between the central and peripheral compartments. Now, there is no loss of substance from the central compartment. When the substance attains an equilibrium level in two compartments, then its rate of loss from the blood is reduced.

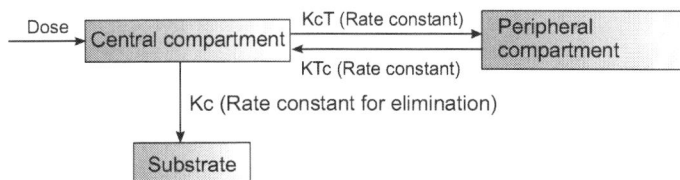

Figure 9.1 Fate of intravenously administered substance in two-compartment model

The toxic effects exerted by the environmental pollutants are mainly due to their chronic exposure into the body of organisms in small quantities. The uptake of small amounts of substances at frequent intervals more or less corresponds to the intravenous administration. In both cases, the plasma concentration will reach a plateau at which the rate of elimination equals the rate of absorption. At a steady state, the rate of entry of a substance into the body exactly equals the rate at which it is removed by biotransformation and excretion. To bring about toxic effect, the toxicant interacts with the binding sites present on the proteins. When the concentration of these sites exceeds certain limits, all the sites will be saturated and will result in decreased elimination rate and prolonged effect. When the toxic substance binds reversibly to active sites, then the extent of toxicity depends on the plasma concentration of the substance. This is because, the magnitude of the effect is determined by the concentration of the substance at the site of action. In general, the highly reactive intermediates exhibit irreversible binding. Here, the intensity of the toxic effect is proportional to the number of target molecules and reactive intermediates especially the total amount of covalently bound reactive intermediates.

Graded and Quantal Responses

The organisms exhibit two types of responds to the toxic substances. If the response in organisms is measurable continuously, then it is **graded** or **gradual response**. If the response is measured by counting the number of individuals who respond to the chemical in a group of affected organisms, then it is called **quantal response**. The intensity of the effect or response caused by a substance generally depends on the quantity or concentration or dose of the administered substance.

Graded Response in Isolated Organs and Tissues

The toxic effect of a substance can be explained at molecular level based on the interactions between the molecules of the substance and the macromolecules of the organism. These interactions are highly specific, i.e., the toxic substance and the molecular sites of action have particular affinity with each other and these sites of action are called as **receptors**. The interactions between the toxic substances and the receptors include electrostatic interactions, van der Waal's forces, covalent bondings and hydrophobic interactions. In addition, the interactions have been found to involve non-specific receptors also.

The relationship between the substance and the effect in isolated parts of an organism is highly useful to study the mechanism of action of toxicants. When an isolated organ is exposed to various concentrations of the toxicant, a characteristic relationship can be obtained between the concentration and the intensity of the effect. The interaction between agonist and receptor is an association–dissociation equilibrium and can be determined by the dissociation constant (K_D) as shown below.

$$K_D = \frac{[P][X]}{[PX]}$$

where,

[P] is the concentration of non-occupied receptor sites,

[X] is the concentration of free agonist molecules, and

[PX] is the concentration of occupied receptor sites.

The concentration of occupied receptor sites provides stimulus which causes the measurable effect or response. In general, the response is directly proportional to the number of receptor sites occupied by the agonist so that binding of all receptors produce the maximum effect.

Graded Response in Intact Animals

Determination of graded response in intact animals is helpful to examine the toxicity of xenobiotics when the organisms having complex and interrelated systems are exposed to the toxic chemicals. These include alterations in a number of processes such as morphology, histology, physiology and biochemistry of organisms.

Quantal Response

Here, the magnitude of the effect is measured by counting the number of responding individuals in a group of equally exposed ones. The toxicity of a substance is determined from the relationship between concentration and lethality. A group of test organisms are subjected to the same concentration of the substance and the dose is gradually increased in each group. The number of animals that would die within a particular period of the time are scored. To explain the toxicity in quantitative terms, the values of LD_{50} or LC_{50} or ED_{50} are used.

FACTORS AFFECTING MODE OF ACTION

Dose of the substance   The magnitude of toxic effects of a toxicant depends on the **dose** (quantity or concentration) of the substance that enters into the body of organisms. For example, the substances which cause toxicity when taken in micrograms per kilogram of body weight are **extremely toxic**, **very toxic** when taken in milligrams per kilogram, **moderately toxic** when taken in hundreds of milligrams per kilogram, **slightly toxic** when taken in few grams per kilogram and **non-toxic** when taken in five grams or more per kilogram of body weight.

Time-response   It also plays an important role in bringing about a toxic effect. For environmental pollutants and food additives, the toxic impact does not show if they are taken in a single dose but requires a series of doses for a longer period of time. It is also true that a minimum duration is required by the toxic substances to cause toxicity at high concentrations. At the same time, no effect is produced even with a very long exposure period below certain concentrations.

Chemical composition and chemistry of cells   The toxic action of a substance is also related to its chemical composition and the chemistry of the cells of the organs. After entry into the bloodstream, a toxicant is distributed to all the tissues of the body but it accumulates only in some tissues or organs (**selective action**). A toxicant may exert its effect on all protoplasm of an organism or exert a particular effect on a specific tissue without affecting others.

Other factors   The other factors which affect the mode of action of xenobiotics include environmental conditions and health, age, sex and habit of organisms.

MODE OF ACTION OF SELECTIVE TOXICANTS

1. **Chlorinated hydrocarbons** The mode of action of these insecticides on target animals is not well known. It is probable that they could interfere with nerve impulse transmission through neurons and synapses by the way of altering ionic movements.

2. **Organophosphates** These are found to inhibit esterases in living systems and AChE in synaptic nerve transmission.

3. **Carbamates** These insecticides are found to inhibit the action of AChE in insects.

4. **Arsenicals** These are found to interfere with oxidative decarboxylation of α-ketoglutaric acid in the TCA cycle. In combination with glutathion, the trivalent arsenicals form relatively insoluble compounds. In mammals and insects, the arsenic trioxide alters the digestive tract.

5. **Fluorides** As they have a greater affinity to calcium ions, they are found to alter the membrane permeability in organisms causing physiological disturbances.

6. **Nicotine** It is a nervous poison which brings blockage of sympathetic and parasympathetic nerve functions especially paralysis and death.

7. **Plant derivatives** Some plant derivatives like pyrethrum act on the nervous system of insects causing lesions. Rotenone causes muscle paralysis in insects by altering the rate of heart beat. It is also found to disrupt the respiratory metabolism and nerve impulse conduction. Ryanodine is a muscle depressant which interacts with phosphogen and actinomysin reaction.

8. **Rodenticides** Some rodenticides like sodium fluoroacetate influences the central nervous system and

interferes with heart functioning. Some like warfarin acts as an anticoagulant causing internal haemorrhage and death.

METHODS TO STUDY THE MODE OF ACTION OF TOXICANTS

The following are the important methods to study the mode of action of toxic substances.

Radioactive tracer technique   By this method, the toxicants or their analogs are labelled with radioactive elements and are injected into the test organisms. The labelled toxicants are traced to locate the site of accumulation in the body of organisms.

Study of metabolism of toxicants   This method involves the study of transformation of xenobiotics into their metabolites in different tissues, organs or organ systems. As a result, it is possible to know whether the compounds are detoxified to be eliminated or accumulated to bring toxic effects in target sites.

Enzyme studies   The toxicants cause the ultimate toxicity on enzyme or enzyme systems of organisms. Therefore acquiring knowledge about the impact of toxic substances on enzymes finds wide usage in the study of mode of action of biologically active substances.

Studies on symptoms of toxicity   This technique is advantageous in the toxicity responses that are recorded in the vital systems of affected animals by using sophisticated instruments. This instrumentation technology is highly useful to trace the functional disruption of vital systems in the test animals which are affected by xenobiotics. However, the study of mode of action of toxicants requires the simultaneous effects of all the symptoms in the biological system **(syndromes)**.

REVIEW QUESTIONS

1. How can you classify toxicants based on their mode of action?

2. Describe the various theories regarding the mode of action of toxicants.

3. Explain the mode of action of toxicants with the help of compartmental model.

4. What are the factors that affect the mode of action of xenobiotics?

5. Explain the methods in the study of mode of action of toxic chemicals.

6. What are graded and quantal responses? Explain them with examples.

10

TARGET-SITE INTERACTIONS

The manifestation of a toxic effect is due to the interaction between a toxic substance or one of its metabolites and a target molecule in the body of organisms (toxicodynamic phase). All the tissues and organs of the body, to a greater or lesser extent, are susceptible to toxicity but the effects of many toxic chemicals are seen on one or more tissues or organs (target organs of toxicity). In general, the susceptibility depends on one or more factors such as the distribution of a toxic substance in the tissues, metabolic activation of the toxicants within the tissues, physico-chemical nature of the toxicants and the biochemical differences of the tissues. The effects may be hepatotoxic, nephrotoxic, neurotoxic or haematotoxic. However, some substances exhibit multiple-organ toxicity in the sense that they affect several organs at the same time.

ROLE OF TISSUE
DISTRIBUTION OF TOXICANTS

The quantum of any toxic impact depends on the concentration of the substance in the target organ and the ability of the organ to concentrate on a toxic chemical. These processes in turn are influenced by specific active uptake mechanisms, tissue-specific binding processes, physico-chemical concentrations and ionization of the toxicants. For example, the pulmonary epithelium is capable of concentrating and metabolizing a number of toxicants from the peripheral circulation. In kidneys, some chemicals are reabsorbed as low-molecular weight proteins and concentrated in the cells of proximal tubes, from where they are mobilized to a number of organelles. Some chemicals are crystallized in the tubules, as their metabolic products are poorly water-soluble.

ROLE OF METABOLIC ACTIVATION

In general, most of the toxic chemicals are in non-reactive states and they require metabolic activation to bring about toxic effects. This is accomplished either by the formation of chemically stable metabolites which may have an altered activity with respect to the parent compound, or by the formation of chemically reactive metabolites which are capable of interacting with cellular constituents or by the formation of harmful endogenous by-products in excess quantities. The stable intermediates are mostly electrophiles which have a long lifespan. The hard electrophiles react with nucleophilic sites such as S-atom of proteins and N,O and C atoms of the bases of nucleic acids. But they react poorly with glutathione and protein thiols. On the other hand, the soft electrophiles react with thiols and glutathione and not with other nucleophilic sites of biological macromolecules. Usually the liver is not a target organ for toxicity, though it is the major site of metabolism of xenobiotics. This is because many reactive metabolites formed in the liver are stable enough to be transported to other organs in which the toxicity occurs. In some cases, the metabolic activation occurs first in the liver followed by further reactions in the target organs themselves.

ROLE OF PHYSICO-CHEMICAL NATURE OF TOXICANTS

The intensity of toxic effects is closely related to the physico-chemical nature of the toxic substances. For example, the bases of DNA contain nucleophilic groups so that they have a strong affinity to electrophilic substances like alkylating compounds. Therefore, these substances are both mutagenic and carcinogenic as they react with DNA and alter its properties. In the reaction between alkylating substances and nucleophilic groups, the formation of covalent bond is known as nucleophilic substitution. Proteins also have nucleophilic groups with which the electrophiles can form covalent bonds. When the electrophiles react with an enzyme or an enzyme system, the enzyme activity is inhibited with the resultant mutagenicity.

It is also shown that the toxic substances which produce similar type of effects will differ in the intensity of their effects. These differences are due to toxicokinetic behaviour and toxicodynamic properties of the toxicants. The toxicokinetic processes mainly involve diffusion of substances based on concentration gradient across the membrane. Thus, the permeability of the substance determines its distribution over the water phase and membrane. Therefore, the membrane acts as a barrier for hydrophiles, whereas the hydrophobic or lipophilic compounds pass through the membrane more easily. The diffusion in turn depends on the degree of ionization of a toxicant in a specific pH medium.

The binding of the toxicant to plasma proteins occurs through electrostatic interactions in which the plasma proteins have more affinity towards hydrophilic substances.

The nature of the enzymes and its reactivity with the toxic substances determine the electrostatic interactions. If the active site of an enzyme is apolar or hydrophilic, then hydrophobic interactions dominate. If it is more polar, then electrostatic interactions and hydrogen bonding will occur.

RECEPTOR THEORY AND MODE OF INTERACTIONS

In many cases, the toxic effects are organ-specific because a substance or a metabolite may react with specific receptors or certain organs or organ systems and does not affect other receptors. In other words, the toxicants reach the specific target site in the body of organisms through several membrane barriers and interact with certain receptors. This interaction stimulates alterations of normal physiological and biochemical reactions, leading to ultimate toxic effect. The impact of the toxicant is widespread when the interacting receptor controls the functions of most of the cells. The effect will be dangerous, if the receptor is responsible for a vital function. The toxic effect will be more specific and highly dangerous, if the interaction involves a specific receptor responsible for a specific function.

The toxicant receptor can be defined as a macromolecule such as protein, enzyme or nucleotide with which the toxic substances interact to produce toxic impacts. Any functional macromolecule of organisms can act as a toxicant receptor which is mainly found in cell membranes and organelles as well as in cytosol of specific cells. The toxic substance can bind only to particular receptors to exert its effects. Thus a specific complex is formed between the active substance called agonist and the receptor as shown in Figure 10.1.

Agonist   Receptor   Agonist–receptor complex

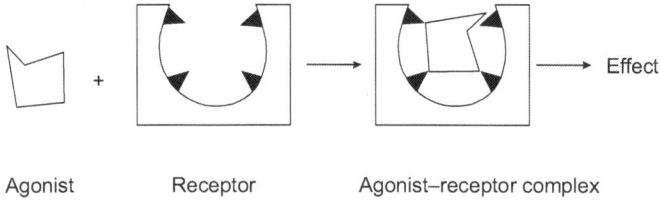

Figure 10.1 Interaction between an agonist and a receptor

The percentage of effect caused by a certain concentration of agonist corresponds to the percentage of receptor sites on the agonist and is determined by the concentration of the agonist at the receptor sites and by the affinity of that receptor for the agonist. It is also known that the binding between an agonist and a receptor site is reversible.

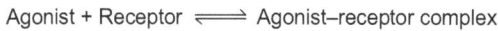

Agonist + Receptor $\rightleftharpoons$ Agonist–receptor complex

If a substance reacts with the receptor sites of macromolecules resulting in the inhibition of the action of the agonist, then it is called as **antagonist**. In other words, the antagonists are the substances which reduce the effect of an agonist. The antagonists are categorized into four types which are as follows:

1. **Competitive antagonist** It is the substance which binds to the receptor as agonist but producing no effects or a smaller effect than the agonist.

2. **Non-competitive antagonist** It is the substance which binds to the receptor sites in such a way that its effect cannot be overcome by the addition of more agonists.

3. **Chemical antagonist** It is the substance which binds to the agonists thus making it inactive.

4. **Functional antagonist** It is the substance which acts as an agonist causing an opposite effect.

NATURE OF TOXICANT–RECEPTOR INTERACTION

The toxicant–receptor interaction mainly depends on the receptor site of macromolecules in the body of organisms and the binding of appropriate toxicants to it. In general, most of the toxic substances have similar structures like that of receptor sites and are capable of tight binding with their respective receptors. The toxicant–receptor binding triggers the operation of the toxic substance leading to physiological, biochemical, neurological, histological and other effects. After producing the effect, the agonist dissociates from the receptor site for further action. In a situation where there is a failure of dissociation of agonist, the disruption on the receptor site is intensified so that the agonist acts as a toxic substance. But the antagonist prevents the binding of agonist with receptor thus inhibiting the effect of agonist as shown in Figure 10.2.

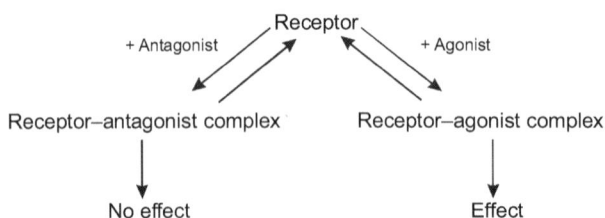

Figure 10.2   Toxicant–receptor interaction

SPECIFIC TARGET-SITE INTERACTIONS

Usually toxic substances are chemically stable and exert their actions by interfering with physiological or biochemical homeostatic mechanisms. However, many adverse effects of toxicants are due to the impairment of normal physiology of organism without causing cell death or cytotoxicity which occurs only when the organisms are exposed to highly harmful toxicants. Many substances can also affect the regulation of cell division. The substances, which bring non-lethal genetic alterations in somatic cells resulting in mutation, constitute carcinogens. More recently, it is also shown that the toxicants, which do not interact directly with the genome, can also produce cancer through epigenetic mechanisms. In this way, the substances may act either through physiological mechanism, or through cytotoxicity, or through producing proliferative lesions.

Effects of Toxic Interactions

Disturbed normal biochemical processes The energy required by the body of organisms is derived by the oxidation of carbohydrates and lipids via the synthesis of ATP molecules through oxidative phosphorylation. As the toxic substances affect the transport of oxygen in the blood, the availability of oxygen to the tissues is diminished and the synthesis of ATP is inhibited. This in turn, not only causes a loss of function but also a marked rise in body temperature. The redox cycling of various toxicants inflict many toxic responses including mutation and cancer as a result of interactions with DNA, membrane damage by lipid peroxidation and biochemical disturbances by enzyme inactivation.

Disturbed excitable membrane function Many toxicants interfere with the closing of sodium channels which result in the depression of central nervous system and thus act as neurotoxins.

Disturbed calcium metabolism Many toxic chemicals disrupt the transport mechanism of calcium ions causing severe effects on individual cells. Higher concentrations of intracellular calcium ions bring alterations in the cytoskeleton, cause leakage of plasma membrane, impair mitochondrial function and activate calcium-dependent enzyme activities.

Covalent bonding   In many cases, the toxic substances bind with the macromolecules of the cell by covalent linkage which is very strong and irreversible. For example, the covalent bonding occurs in the interactions between electrophilic metabolites and nucleophilic substances of DNA molecule, resulting in altered gene function.

Disturbed functioning of membranes   Parts of cellular membranes possess a number of lipophilic groups of molecules which attract the hydrophobic or lipophilic substances. Thus a close affinity between the receptor and the toxic substance is established. By this interaction, the receptor and the toxicant are very closely bound and the uptake of apolar molecules into the cell membranes brings changes in the functioning of membranes.

Disturbed gene expression (Genotoxicity)   The oxidative metabolism of xenobiotics produces electrophilic substances which interact with the nucleophilic sites of DNA. As a result, gene expression is altered. This causes cell death but more frequently somatic mutation, ultimately which lead to cancer. This phenomenon is called genotoxicity. However, it is shown that not all the toxicants which cause cancer are genotoxic. Some substances produce proliferative lesions leading to altered cell division without acting at the genome level. Such effects are considered to be non-genotoxic or epigenetic carcinogenicity.

REVIEW QUESTIONS

1. Analyse the factors which play an important role in target-site interactions.

2. State the role of tissue distribution, the metabolic activation and the physico-chemical nature of toxicants in target-site interactions.

3. Describe the receptor theory with regard to the mode of interaction of toxic substances.

4. Illustrate the nature of toxicant–receptor interaction.

5. Narrate the various specific target-site interactions.

1 1

COMBINED ACTION
OF TOXICANTS

At present, the ever-increasing human needs have resulted in the use of combinations of chemical substances. The introduction of more than one substance at a time into organisms causes combined effects which are qualitatively and quantitatively different from that of the effects of individual substances. Therefore, it is important to evaluate the interaction between two or more substances in a mixture, the consequences of their toxicity and their mode of action in combination. The interaction between more than two substances may be complex and difficult to interpret. While making studies on toxicant mixtures, it is essential to know whether the substances influence each other, in what ways and with what effects they influence each other as well as the extent of the effect.

QUALITATIVE AND QUANTITATIVE
ASPECTS OF TOXICITY OF MIXTURES

The combined action of toxic substances may occur during the exposure phase, kinetic phase, dynamic phase or at their physico-chemical interactions. In combination, the substances can affect the absorption, protein binding, biotransformation and excretion of other substances. The combined action of substances with opposite effects depends on their dose and duration. In both the environment and the food, the physico-chemical interactions between the substances could produce new compounds which would exhibit entirely different toxic effects than the original substances. In addition, some substances react with each other to produce a compound which may be toxic or harmless.

In a mixture of toxicants, the concentration or dose of a substance A is expressed in toxic unit (T_A). As the dose (D_A) is related to the effective dose (ED_A) of the substance, the toxic unit (T_A) is given by

$$T_A = D_A / ED_A$$

The mixture of dose represents the sum of the numbers of toxic units of the substances in a mixture and is represented as M. For a mixture of two substances, the effects of combination are compared by taking both quantities as effective doses or toxic units. Thus

$$T_A = D_A / ED_A \text{ and } T_B = D_B / ED_B$$
$$M = T_A + T_B$$

TYPES OF COMBINED ACTIONS

When the combined effect of substances A and B is equal to the sum of the individual effects of A and B, then it is called **additivity**. The counteraction of two substances is called **antagonism**. This term can also be used for situations in which one non-toxic substance suppresses the action of another toxic substance. Antagonism may also prevail when the effect of one substance is reduced by the second one in which the first substance will not reduce the effect of the second one. When a substance, which has no ability to produce a toxic effect, induces the action of another toxicant, then it is called **potentiation**. When the substances in a mixture produce a greater effect than the sum of the effects of the individual substances, then it is called **synergism**. Sometimes, one substance may increase the effect of another one in a mixture of toxicants which may also be referred to as synergism. In nature, only a very few substances are shown to bring about synergistic action.

Hewlett and Packard have distinguished the combined action of toxicants into four types.

Simple similar action Here, the substances in a mixture have the same mode of action and do not influence each other's activity.

Independent action Here, the substances in a mixture have different modes of action but do not influence each other's activity.

Complex similar action and dependent action These two types of action occur when there are interactions between the different substances in a mixture. This means that the substances do not exert their effects independently. The combined toxicity of these types may range from partially additive to potentiation and antagonism.

According to Chou, the combined action of substances may either be mutually exclusive or mutually non-exclusive. When the action of a substance is entirely independent of another substance in the mixture, then it is called **mutually exclusive**. When the substances of a mixture possess mutual influence on their action, then the action is called **mutually non-exclusive**.

DETERMINATION OF COMBINED TOXICITY

The nature of the combined interactions of the substances in a mixture of toxicants can be evaluated by adopting various methods.

Combination Index (CI) Method

The combination index value (CI value) for a mixture of two substances can be calculated by using the following formula:

$$CI = D_A / D(x)_A + D_B / D(x)_B + \alpha D_A D_B / D(x)_A \cdot D(x)_B$$

where,

D_A = the dose of substance A in the mixture that has X% effect

D_B = the dose of substance B in the mixture that has X% effect

$D(X)_A$ = the dose of substance A which caused X% effect

$D(X)_B$ = the dose of substance B which caused X% effect

$\alpha = 0$ or 1

$X = 0.5$ and $D(X) = LD_{50}$

If the CI value is 1, then the reaction is additive; if lesser than 1, it is synergistic and if greater than 1, it is antagonistic.

Co-toxicity Coefficient Value (CTC value) Method

The co-toxicity coefficient value (CTC value) for a mixture of two toxicants A and B can be calculated by the formula:

$$CTC\,value = \frac{Actual\,toxicity\,indices\,of\,AB(TI_{AB})}{Theoretical\,toxic\,indices\,of\,AB\,(TI_{AB})} \times 100$$

$$\text{Actual TI}_{AB} = \frac{LC_{50} \text{ of A}}{LC_{50} \text{ of AB}} \times 100$$

Theoretical TI$_{AB}$ = (TI of A X% A in AB) + (TI of B X% B in AB).

If the CTC value is more than 100, the action is synergistic and it is antagonistic when it is less than 100.

Index 'V' Value Method

The calculation of *V* values for the toxicants A and B in a mixture can be done by using the following formula:

$$V = \frac{\frac{1}{2} LC_{50}(A) + \frac{1}{2} LC_{50}(B)}{LC_{50}(A+B)}$$

If the *V* value is < 0.8, then the combined action will be antagonistic. If it is in between 0.8 to 1.5, the action will be additive. If the value is 1.5 or more, then it is potentiation.

Mixture Toxicity Index (MTI) Method

This method is useful to assess the nature of combined toxicity involving large numbers of substances and the index can be derived by applying the following formula:

$$MTI = (\log M_0 - \log M) / \log M_0$$

where,

M = Total number of toxic units and

M_0 = The range of values from 0.

If the MTI value is <0, then the combined action is antagonism; if it is equal to 0, then it indicates no additivity; if it is 0–1, it shows partial additivity; if it is equal to 1, it is concentration additivity and if it is >1, it is potentiation.

Linear S Index Method

Linear S index is calculated by the formula:

$$S = \left[\frac{Am}{Ai}\right] + \left[\frac{Bm}{Bi}\right]$$

where,

A and B = Individual pollutants

i = LC_{50} of an individual pollutant

m = LC_{50} of a pollutant mixture

S =   Sum of biological activity

This index quantitatively expresses the toxicity in which the action of mixtures of pollutants will be additive if the S value is >1 and more than additive if the S value is < 1.

COMBINED ACTION OF TOXICANTS IN AQUATIC ENVIRONMENT

While mixtures of two or three substances exert different types of combined effects, a mixture of many different substances in an aquatic environment usually causes additivity or partial additivity. The partial additivity not only leads to the mortality of aquatic organisms but also produces sublethal effects in them.

The static bioassays carried out in the laboratory by using different kinds of effluents, namely, tannery effluent, paper and pulp mill effluent, and dyeing factory effluent on the dragonfly larvae reveal that the tannery effluent has more toxic effect followed by the paper and pulp mill, and dyeing factory effluents. When the effluents are mixed at a definite ratio in different combinations, the toxicity is found to be

Table 11.1 Individual and combined toxicity of different effluents on the dragonfly larvae

Mixture	Ratio of combination	LC$_{50}$ 96 hr value(%)	95% Confidence limits	
			Lower	Upper
Tannery effluent (TE)		5.00	2.49	10.05
Paper and pulp mill effluent (PE)		44.00	26.02	74.40
Dyeing factory effluent (DE)		60.00	20.44	176.04
TE : PE : DE	1 : 1 : 1	43.50	16.50	114.01
TE : PE	1 : 1	12.50	4.61	33.88
DE : PE	1 : 1	52.00	28.76	94.01
TE : DE	1 : 1	12.40	6.60	23.23

higher in combinations when compared to the effect of individual effluents (Table 11.1). This could be due to the manifestation of additive effect caused by the interaction of the effluents resulting in potentiation of toxicity.

The combination of paper and pulp mill effluent and dyeing factory effluent causes a toxic effect indicate synergism. The same is also true for the combined action of paper and pulp mill effluent and tannery effluent and is supported by their cotoxicity coefficient values. The index V values for the combinations of paper and pulp mill effluent and tannery effluent as well as tannery effluent and dyeing factory effluent show the prevalence of potentiation whereas the combination of dyeing factory effluent and paper and pulp mill effluent, the additive nature (Table 11.2). Again the S values for the above mixtures indicate additive toxicity.

Table 11.2　Combined action of various kinds of effluents on the dragonfly larvae

Index values	Treatment		
	PE : TE (1 : 1)	DE : PE (1 : 1)	TE : DE (1 : 1)
Co-toxicity coefficient values	224.00	101.95	48.17
Index V values	1.96	1.00	2.60
S values	0.58	1.98	2.69

Table 11.3   Combined action of two kinds of effluents on the dragonfly larvae

Treatment	Ratio of combinations	Co-toxicity coefficient	*V* values	*S* values
PE and TE	1.00 : 1.65	57.73	2.53	0.98
PE and TE	3.50 : 1.00	23.44	1.19	2.07
PE and TE	1.00 : 2.50	26.10	1.67	1.48
PE and TE	1.50 : 1.00	81.67	2.78	0.89
PE and TE	1.00 : 2.00	40.66	2.09	1.18
PE and TE	1.35 : 1.00	94.23	3.09	0.80

On the other hand, the combined toxicity of tannery effluent and paper and pulp mill effluent at different ratio of combinations envisages the occurrence of antagonism which is supported by the cotoxicity coefficient values which are lesser than 100 and by the index *V* values which are greater than 0.8 (Table 11.3). The above mixture in certain ratios has the *S* values less than 1 indicating the presence of lesser than additive toxicity.

REVIEW QUESTIONS

1. Explain the necessity for evaluating the combined action of toxicants.

2. Define and narrate different types of combined actions of xenobiotics.

3. How can you determine the combined toxicity in a mixture of pollutants?

4. With experimental evidences, explain the phenomena of synergism and antagonism.

12

DOSE–RESPONSE RELATIONSHIP

The quantity of toxicant received by an animal at a given period is called **dose** which is expressed in mg, mol, ml, etc. per kilogram body weight of the animal. In acute toxicity studies, the toxic response is observed in a concentration which will cause mortality of 50% test organisms and at two other concentrations, one of which causes more than 50% mortality and another less than 50% mortality. In chronic toxicity studies, the toxicity is to be assessed at least in three concentrations in which the highest concentration would cause toxic effects and the lowest concentration would not cause toxic effects. Thus the magnitude of harmful effects due to different doses of toxicant on test animals is referred to as **toxic response.** In addition, the duration of exposure is also important and it depends on the species and its developmental stage, age, sex, etc. In general, the duration of exposure is from 48–96 hrs in acute toxicity tests whereas it is more than three weeks in chronic toxicity tests.

The term **effect** implies a biological change in animals whereas the term **response** indicates the proportion of the population which exhibits a defined effect. The acute effects (mortality) occur rapidly due to short-term exposure of organisms to a toxicant and chronic effects (lethal or sublethal such as behavioural, physiological, biochemical and histological changes) are produced due to the long-term exposure of organisms to a toxicant.

The extent of toxicity of a substance which produces the toxic effect is mainly due to its dose. However, an exact quantification of the intensity of toxicity is not possible. A numerical approach can be made in which the arbitrarily chosen aspects such as mortality percentage, functional damages to organs, development of tumours, etc. can be studied. In this way, the numerical method provides possibilities to study the intensity of the toxic effects accurately. Each and every dose or concentration exhibits a specific frequency with which an effect is produced in a group of organisms. So, the frequency of a toxicant is directly proportional to its dose and can be expressed graphically in the form of a **frequency curve**. This graphic extrapolation also helps to know the median value of the effective dose as well as non-effective dose. The no-effect level is important for evaluating safety standards for toxic substances.

The graphic representation which shows the relationship between dose and effect of a toxicant on a test animal is the dose–response curve as shown in Figure 12.1. When the dose of a chemical is plotted logarithmically on a graph sheet, a sigmoid curve is obtained and is useful to estimate LC_{50} or LD_{50} value of a substance. The range between the dose A and B on the graph represents the **non-effective** or **non-lethal dose**. The point B denotes LD_0 which is determined by the presence of a small number of relatively sensitive animals. The point C denotes LD_{100} which is determined by least-sensitive animals. Therefore, to find out the lethal effect of a substance the determination of LC_{50} and LD_{50} is preferred to that of LD_0 and LD_{100}.

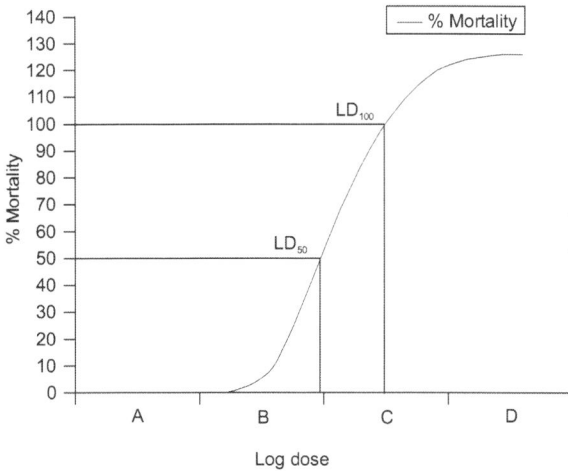

Figure 12. 1   Dose–Response curve

RESPONSES TO TOXICANTS

If the response is measurable on a continuous scale, then it is **graded** or **gradual** or **quantitative response**. This can be observed in organ or tissue preparations or in intact animals. The exposure of an isolated organ or an intact organism to a toxicant is useful to know the mechanism of action of the substance. When an intact organism is used, a number of effects such as changes in growth rate, organ weight, nature of blood pressure, level of biochemical constituents, etc. can be studied. In a single individual, a noticeable effect could occur only when a **threshold dose** of a toxicant has been exceeded. The threshold dose can be calculated from the dose–response relationship and is useful to determine the no-effect level of a toxicant. However, most of the toxicological studies are not concerned with gradual responses.

When the intensity of the effect is measured by counting the number of individuals in a group which are equally exposed to the toxicants, then it is called as **quantal response**. The relationship between dose and lethality is an example of this response in which the animals are randomly distributed in groups and the dosage is gradually increased. When all the test animals are equally sensitive, there will be a threshold dose below which all animals would be alive and above which all would die.

EFFECT OF DOSE ON EXPOSURE TIME

The relationship between time and dose is also important in bringing out the response by a toxicant. The factor time has two important aspects, namely, time until the effect and duration of the effect. However, the duration of the effect is applicable only when the effect is reversible. Generally, a single dose of a substance does not bring out the effect but requires a series of doses for a longer period of time. This is the **cumulative dose** which is directly proportional to concentration times and the response produced is called **cumulative response.** An effect requires a minimum period of time to bring action. Even with a very long exposure period, no effect manifests below a certain concentration level. This concentration is called **incipient effective concentration**. When the response is 50% mortality, then it is **incipient** LC_{50}. The time between exposure (single dose or short-term exposure) and effects is called **latency period** and it depends on the dose of the toxicants.

REVIEW QUESTIONS

1. Define the terms dose and toxic response with examples.
2. How can we relate dose and toxic response by graphical method?
3. Explain the graded and quantal responses with suitable examples.
4. How does the duration of exposure alter the effect of dose?

13

TERATOGENS

In a broad sense, teratology involves any abnormality which occurs in the course of development of embryo or foetus. Therefore, it deals with the study of **congenital malformations** or **developmental malformations** which appear during the development of foetus. The term **teratogenesis** refers to the formation of congenital defects due to the interference of toxic substances with the normal development of either embryo or foetus causing morbidity and mortality in newborns. The induction of foetal malformations by the toxicants is called **teratogenicity**. The **teratogens** can be defined as the toxicants which induce abnormal development in foetuses.

CAUSES OF TERATOGENESIS

It is found that nearly 10% of malformations in man are due to the environmental factors, 10% due to hereditary factors and 80% due to the interaction of genetic and environmental factors. At present, it is well established that various environmental toxicants, viruses, nutritional and oxygen deficiencies, etc. are the major causatives of teratogenesis.

Chemical Agents

Many natural toxins like nicotine; solvents like benzene, carbon tetrachloride, dimethylsulphoxide, propylene, xylene, glycol, etc. insecticides; herbicides; fungicides; azodyes like Trypan blue, Evan's blue, etc., antibiotics like penicillin, streptomycin, tetracycline, etc. and many drugs and chemicals like caffeine, carbutamide, chloropromazine, hydroxyurea, nitrosamines, quinine, rauwolfia, alkaloids, etc. are found to produce many abnormalities in human foetus.

Radiation

Many radioactive elements cause teratogenesis depending on the doses of radium and the stage of development of the affected victim. When large doses of radiation are administered to pregnant women, defects like microcephali, blindness, cleft palate, etc. are found to occur in foetuses. Among the Japanese women who were pregnant during atomic bomb explosions, 28% were found to be aborted and 25% to give birth to babies who died in the first year, and about 25% of the children were found to have abnormalities in their central nervous system.

Hypoxia and Vitamin Deficiency

In many experimental animals, the hypoxic condition is found to cause malformations during embryonic development. It is shown that the children born at high altitudes are smaller with light weight than those who are born at sea level. In addition, nutritional deficiency, mainly vitamin deficiency, is also reported to have teratogenic effect.

Viruses and Bacteria

In nature, a variety of viruses are found to cause congenital malformations in human beings. For example, Rubella virus causes malformations of eye and internal ear as well as mental retardation. Herpes simplex virus also causes mental retardation, microcephali and eye defects in the child. Syphilis is shown to cause deafness and mental retardation in newborns due to the infection of the bacteria *Treponema pallidum*.

Hereditary Factors

Many chromosomal aberrations are found to cause teratogenesis in human beings for example, the occurrence of trisomic condition which results in autosomal abnormalities such as Down's syndrome and of allosomal abnormalities like Klinefelter's syndrome and Turner's syndrome. It is reported that certain abnormalities (cleft lip or cleft palate) are inherited in the Mendelian ratio as recessive characters.

Sensitivity of the Foetus to Teratogens

During pre-differentiation stage, the embryonic cells multiply and differentiate at high rates so that the embryo is more susceptible to teratogenic agents which will either cause death of the embryo or produce no apparent effect on the embryo (5–9 days include resistance stage). In the embryonic stage, the cells undergo differentiation, mobilization and organization including organogenesis. As a result, the embryo becomes more susceptible to teratogens (10–14 days of gestation in rodents and 14th week in humans). However, not all organs are susceptible in the same period of pregnancy. For example, the teratogen treatment on 10th day of gestation in rats causes 35%, 33%, 24%, 18%, 6% and 0% defects in the brain, eye, heart, skeletal, urinogenital and palate respectively. The foetal stage is characterized by growth and functional maturation, thus the teratogens do not cause morphological defects but may cause functional abnormalities.

MODE OF ACTION OF TERATOGENS

There are many mechanisms by which the teratogens bring out malformations either independently or in combinations.

Interference with Nucleic Acids

Many teratogenic agents like alkylating agents, antimetabolites, intercalating agents, etc. interfere with nucleic acids at replication, transcription and translation levels thus arresting the DNA synthesis.

Mitotic Interference

The teratogens sometimes interfere with spindle formation during mitotic division with incorrect separation of chromatids leading to chromosomal aberrations.

Inhibition of Enzymes

The inhibition of certain enzymes is shown to induce malformation by interfering with differentiation of growth of the embryo. For example, 5-fluorouracil inhibits thymidylate synthetase activity; 6-amino nicotinamide inhibits the activity of G 6-phosphate dehydrogenase, folate antagonists inhibit the activity of dihydrofolatereductase, and so on.

Deficiency of Energy Supply and Osmolarity

Some teratogens affect the energy supply in metabolic pathways by restricting the availability of substrates either directly (dietary deficiencies) or through analogs or antagonists of vitamins, essential amino acids, and so on. Either hypoxia or agents which induce hypoxia such as CO and CO_2 bring out teratogenic effects by inhibiting the metabolic processes through oxygen deficiency and osmolar imbalances. However, the mode of action of many teratogens is still uncertain.

EVALUATION OF TERATOGENICITY

In evaluating teratogenicity, *in vivo* tests such as cell culture, organ culture, etc. are not put to routine use. However, they are useful as screening procedures to locate the target organs and to know the mode of action of teratogens. Only poor knowledge is available with regard to the mechanism of teratogenesis and differential response in animal species. Therefore, the incidence as well as the intensity of teratogenic effects should be considered in teratogenic tests.

Test Animals

Test animals must practically be chosen based on a balanced comparison with the physiology of man. Therefore, the animals should be young, mature and healthy with a relatively short lifespan and large litter size. Animals like rats and rabbits are commonly used because of their ready availability, easy handling, litter size and short gestation period. Animals like dogs, cats and pigs are also used. Primates are suggested to be used in teratogenic tests as they are phylogenetically related to human beings.

Administration of Teratogens

In routine teratogenic tests, the teratogen is administered to the pregnant test animals at the time of embryonic organogenesis during which the embryo is more susceptible. Usually the test samples are administered through the route which is found favourable for human exposure. In general, three or four doses are selected to be administered along with the control which is given physiological saline for stipulated time. At the end of the experimental period, the observations are made on the pregnant mother animals and foetuses.

Examination of Pregnant Animals

The mother animals are examined daily for overall toxicity and many of them show signs of abortion or premature delivery in the form of vaginal bleeding. Such animals are killed before the young ones are born for *in utero* inspection. The food intake, weight, weight increase and general health of the animals are also recorded to know the direct and indirect embryonic foetal toxicity.

Examination of Foetuses

A day before the expected delivery, the foetuses are removed from the mother and are examined for a number of developmental disorders like death, abnormal weight and malformations, and for skeletal abnormalities (Alizarin red stain method) and visceral defects (microtechnique). The uterus of the mother is also observed for the number of corpora lutea, number and position of implantations, resorptions, and so on.

Prenatal and Post-natal Exposure

Here the mother animals are treated during the prenatal (starting from day 15th of gestation) and post-natal period (birth and lactation period). After this period, the young ones are killed and examined for survivability, growth retardation, malformations and a number of functional disorders.

Data Analyses and Extrapolation to Human Beings

The qualitative data obtained from the experimental animals must be combined with the quantitative factors. In other words, the effects of higher doses must be extrapolated to lower doses also in order to know the no-observed-adverse-effect level in animals which is then extrapolated to the exposure level in relation to humans. If such data are not available, the data of teratogenicity are combined with those of other toxicity studies to get the final evaluation. The developmental defects can be prevented by establishing the cause of a defect and by taking measures to prevent or reduce the exposure to that particular substance as well as by restricting the use of those drugs during pregnancy.

REVIEW QUESTIONS

1. Define "teratogen" and "teratogenicity".
2. Write an essay on teratogenic agents.
3. Describe the mechanisms by which the teratogens act.
4. Explain the procedures to evaluate teratogenicity in animals.

1 4

CARCINOGENS

Cancer is a term used to denote a wide variety of malignant, autonomous growth of tissues (neoplasia) causing deleterious effects due to their invasive and metastasing characters.

NEOPLASIA

The term **tumour** refers to an actively growing tissue in which the normal growth controlling mechanism is completely impaired resulting in an uncontrolled autonomous growth. The formation of tumours is called **neoplasia**. Neoplasia may either be **hyperplasia** (an increase in number of cells), **hypertrophy** (an increase in cell dimensions and not in numbers), **metaplasia** (particular cell types in tissues or organs change into other types), or **anaplasia** (change in cellular organization).

Based on the biological behaviour, neoplasia are classified into two types.

i.   **Benign neoplasia** in which the cells grow relatively slowly and are well defined, encapsulated, non-invasive and well differentiated.

ii. **Malignant neoplasia** in which the cells grow rapidly and display abundant mitosis and are less well-defined, not well-encapsulated, invasive and relatively undifferentiated. The malignant cells have high proliferative index, altered ultrastructural changes, enlarged size and larger nuclei.

CLASSIFICATION OF CARCINOGENS

The agents which cause cancer are called as **carcinogens** and the phenomenon of cancer formation is called as **carcinogenesis**. Carcinogens are classified in a variety of ways which are listed in the following.

Classification Based on Functional Aspects

Based on functional aspects, the carcinogens are of the following types.

Procarcinogens   The agents from which the ultimate carcinogens are formed through one or more biotransformations.

Proximate carcinogens   They form the substrate for enzymatic or non-enzymatic reactions to form the ultimate carcinogens.

Ultimate carcinogens   These are the reactive molecules which react with cellular macromolecules initiating carcinogenesis.

Synergistic carcinogens   They enhance the activity of another carcinogen in combination.

Co-carcinogens   This is an enhancing substance but is not carcinogenic.

Anticarcinogens   They reduce the activity of another carcinogen.

Classification Based on Mode of Action

Based on their mode of action, the carcinogens are categorized into two types.

1. **Genotoxic carcinogens** These are procarcinogenic agents which initiate carcinogenesis (initiators). These interfere with DNA so that the genetic information is altered in cells resulting in the loss of normal control over the cell division. Therefore, the cells acquire relative autonomy with regard to cell division, e.g. mustard gas, nitrosamines, aromatic amines, benzopyrene, polycyclic and heterocyclic aromatic hydrocarbons, halogenated hydrocarbons, hydrazines, formaldehyde, metal compounds such as Ni, Cr, Pb, Co, Ca, Ar, etc.

2. **Epigenetic carcinogens** These stimulate cell division (promoters) and act as mitogenic agents (exogenic and endogenic), cytotoxic agents (specific and non-specific) or immunosuppressive agents, e.g. hormones, phenobarbital, carbontetrachloride, clofibrate, cyclophosphamide, etc.

Classification Based on Origin

Based on their origin, the carcinogens are grouped into two types.

1. **Synthetic or man-made carcinogens** These include a number of environmental toxins, socio-cultural substances and drugs.

2. **Natural carcinogens** These include microbial substances, substances derived from plants and animals, naturally occurring radiations and metals.

Classification Based on Potency

Based on the strength or potency, the carcinogens are classified into three types.

1. **Proved carcinogens** These are the carcinogens whose activity has been established in epidemiological studies.
2. **Suspected carcinogens** These are suggested carcinogens but not with conclusive evidence.
3. **Potential carcinogens** These have similar chemical structure like those of proved carcinogens but are not yet tested.

ENVIRONMENTAL CARCINOGENS

A large number of chemical and biological toxins are found to be carcinogenic and are found in air, water, soil, food and drugs. The polyaromatic hydrocarbons (PAH) are the major carcinogens found in air. Water contains PAH, pesticides, asbestos, vinylchloride, etc. Soil contains PAH and pesticides. Food contains pyrolysed protein-derived amino acids, aflatoxins, nitrosamines, safrole, azo dyes, metanil yellow, saccharine, etc. In addition, the lifestyle practices like tobacco chewing, cigarette smoking and drinking give place for carcinogens like CO, hydrazine, vinylchloride, nitrosamines, etc.

Naturally occurring carcinogens include physical agents such as non-ionizing radiations and ionizing radiations. The chemical agents include both inorganic metals and non-metals and organic agents such as cancer-causing viruses and low-molecular weight chemicals from bacteria, yeast, fungi, plant and animal products. The major carcinogenic drugs are neuroleptic agents, clofibrate, phenacetin, phenobarbitone, cyclophosphamide, arsenic, metronidazole, griseofulvin, etc.

MODE OF ACTION OF CARCINOGENS (CARCINOGENESIS)

The carcinogenesis or the development of cancer involves three phases, namely, **initiation phase** in which the cells are exposed to a carcinogen so that the daughter cells acquire autonomic cell division, **promotion phase** in which the cells are stimulated to divide and become detectable neoplasms and **progression phase** in which the tumour increasingly damages the host and finally destroys it.

The chemical carcinogens have been well studied. Chemical carcinogenesis is a multistage process in which a genotoxic carcinogen interacts with the DNA of the cell which either dies or becomes a normal cell through DNA repair. If these two processes do not occur, then the affected cell may remain dormant for a long period of time before it becomes a tumour. Even though carcinogens also interact with macromolecules other than DNA, such as RNA, proteins and lipids, DNA is the primary target. Both physical (ionizing radiation) and chemical carcinogens as well as biological carcinogens interfere with DNA.

There are different views with reference to the mechanism of carcinogenesis emphasizing various factors which induce cancer.

    i.   *Oncogenes* The oncogenes or "tumour-producing genes" are proved to be present not only in retroviruses but also in normal cells. The carcinogens activate the cellular oncogenes which induce alterations in the expression of proto-oncogenes leading to tumour development. Even the cellular oncogenes in normal cells, on activation, can produce tumour development.

ii. *Suppressor genes and regulatory nucleotide sequences* The carcinogens also suppress the action of suppressor gene and the regulatory nucleotide sequence, leading to malignancy.

iii. *Growth-promoting factors* An enzyme protein kinase C (PKC) found in the plasma membrane has regulatory functions in cellular growth and differentiation. The activation of oncogenes disturbs the functions of PKC, causing cancer.

MECHANISM OF CHEMICAL CARCINOGENESIS

Most of the chemical carcinogens (procarcinogens) are metabolically activated to form reactive electrophiles which bind covalently to the nucleophilic binding sites of macromolecules in the cell including DNA.

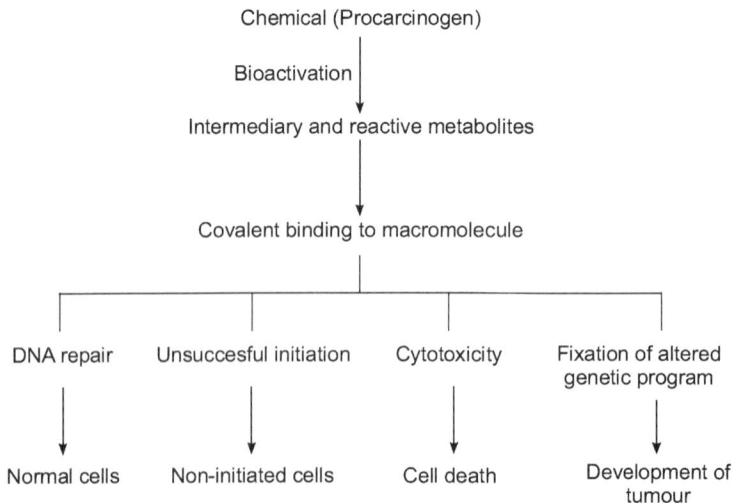

Chemical (Procarcinogen)

Bioactivation

Intermediary and reactive metabolites

Covalent binding to macromolecule

DNA repair	Unsuccesful initiation	Cytotoxicity	Fixation of altered genetic program
Normal cells	Non-initiated cells	Cell death	Development of tumour

During this process, the repair mechanisms of DNA may destroy the covalent binding, so the cells remain normal. Another possibility is that the covalent binding may remain but at the sites of DNA which are not important for the initiation of carcinogenesis so that the cells are non-initiated. Thirdly, if the carcinogen is a toxic one, the cells die without the expression of initiation effect due to a massive interaction of covalent binding. Lastly, the promotion of tumour is initiated in daughter cells through the replication of altered genetic material by the covalent binding. The entire mechanism is shown in the following.

EVALUATION OF CARCINOGENICITY (CARCINOGENICITY TESTS)

Carcinogenicity tests are made in order to get information on the carcinogenic effects of chemicals on the test animals. Rats and mice are usually preferred as test animals in carcinogencity tests because of their small size, short lifespan, ready availability and abundance of information on their response to carcinogens. Hamsters are used in many limited carcinogenicity tests. Dogs and non-human primates are also occasionally used. While both sexes of animals are to be included in the tests, highly inbred strains should not be used.

Duration of the Study

The **short-term tests** include limited carcinogenicity tests and are in general used to study the mutagenesis such as DNA damage and repair, chromosomal aberrations and mutations, cell transformations, and skin, pulmonary, liver and breast cancer in rodents. The **long-term tests** include lifetime cancer studies in rodents and are used to know the presence of substances which are suspected to be carcinogens to which humans are exposed. In rodents, the duration of the study is generally from 18 to 24 months which may also be extended to 24 to 30 months.

Dosage and Route of Administration

More than three doses are selected in order to control in such a way that the dose should produce only some minor effects without affecting the lifespan of test animals. In other words, the selected dose should permit the animals to survive until tumour develops.

The main routes of administration of carcinogens are the inhalation and oral route for industrial and environmental toxicants. The dermal administration is done to detect the local carcinogenicity caused by industrial chemicals and drugs. But intravenous, intraperitoneal and subcutaneous administrations are employed very rarely.

Examination of Test Animals

A group of test organisms are exposed to different dosages of the carcinogen for a particular experimental period and the survival and mortality rates are known. The dead and dying animals should be subjected to massive autopsy and at the expiry of experimental period the surviving animals are sacrificed and examined. Various vital tissues and organs from digestive, skeletal, nervous, circulatory, respiratory, excretory, reproductive and endocrine systems are preserved for histological examinations.

Histochemical Tests

Histochemically the degranulation of ribosomes from rough-surfaced endoplasmic reticulum can be traced, as the carcinogens cause structural disorganization of endoplasmic recticulum in the cells.

Biochemical Tests

As carcinogenesis involves three major steps, namely, initiation, promotion and tumour development, these can be tested by using biochemical markers.

Initiation markers　The initiation of carcinogenesis is due to the covalent bonds between the carcinogenic electrophiles or their active metabolites and DNA. This binding can be qualitatively and quantitatively determined by using **radiolabelled carcinogens** and also by **radioimmunoassay (RIA)**. Moreover, the covalent binding is followed by DNA repair which can be determined by radiolabelled thymidine.

Promoter markers　The promoter substance is administered orally to the test animals after initiation by an initiator such as diethylnitrosamine over a number of months. Now liver neoplasia occurs in which various biochemical alterations such as increased plasminogen activator, increased prostaglandin, γ-glutamyl transpeptidase, etc. can be studied. In addition, certain enzymes like glucose 6-phosphate, ATPase, etc. decrease the condition of liver tumour. Therefore, this test is useful for identifying the impact of drugs, pesticides and antioxidants.

Tumour markers　α- ketoprotein (a product of foetal liver and human chorionic gonadotropin) is a major clinical tool for detecting tumour. Abnormal levels of carcino-embryonic antigen, ferritin and C-reactive protein are indicative of mammary cancer.

Susceptibility markers　In the conversion of carcinogens into reactive metabolites, cytochrome P-450 plays an important role so that the activity levels of these enzymes can be measured. Similarly, in detoxicating reactions, the conjugating enzymes are involved and the activity levels of these enzymes can be measured.

MOLECULAR TECHNIQUES

These include the use of integrated techniques of analytical chemistry, biochemistry, molecular biology and epidemiology to study the exposure of carcinogens, their metabolism and DNA repair. The genetic and cytogenetic markers include chromosomal aberrations including restriction fragment length polymorphism (RFLP), loss of heterozygosity and translocation, sister chromatid exchanges, etc.

BIOLOGICAL MONITORING

This involves the estimation and evaluation of exposed carcinogens in individual animals and helps to know the internal dose of a substance and the differences in absorption, bioavailability, excretion, DNA repair, and so on. The internal dose can be quantified by chemical analyses of the substance in blood, urine and expired air. The biological effective dose of a carcinogen is the one which actually enters the site of action and binds with the macromolecules of the cells.

REVIEW QUESTIONS

1. Define "carcinogen" and "carcinogenicity".
2. On what basis and how are cancer and carcinogens classified?

3. Discuss the mode of action of carcinogens.

4. How can the carcinogenic effects of chemicals be evaluated on test animals?

5. Write an essay on biochemical markers to test carcinogenicity.

15

MUTAGENS

The qualitative and quantitative changes either in individual genes or in the genome are called **mutations** which bring unexpected and undirected changes in the composition of DNA. The mutations may either be spontaneous or be induced by external physico-chemical factors such as temperature, toxic chemicals, ultraviolet rays, radiations, etc. They occur at cellular level either in somatic cells leading to teratogenicity in foetuses and carcinogenicity in adults or in reproductive cells leading to abnormal genetic inheritance in the offsprings. Severe mutations are lethal and so are not carried over to future generations. In the early stages, the cells attempt to repair the modifications in the DNA through DNA repair mechanisms. If the alterations are fixed, then they are irreversible and remain in the population.

TYPES OF MUTATIONS

In a broader sense, mutations can be classified into three groups, namely, **single-point mutation** or **gene mutation** in which small changes occur in the DNA molecule at the level of the bases, **structural chromosomal aberrations** in which major changes occur in the structure of chromosomes and **genomic mutations** in which abnormal number of chromosomes occur (aneuploidy).

MUTATION-CAUSING AGENTS

In the present-day scenario, a variety of toxicants are released into the environment causing an increased rate of mutation. The agents which increase the rate of mutation are called **mutagens** or **mutagenic agents.** The ability of substances to bring out alterations in the genetic material is mutagenicity. The level of mutation is dependent on the dose of these mutagens. The mild expressions of mutations occur only at low concentrations of mutagens. The mutation rates are increased by the following groups of mutagens.

Ionizing radiation This includes X-rays, γ-rays and decayed particles of radioactive elements such as radium, thorium, etc. These mutagens lead to the formation of chemically active ions inside cells which break DNA strands causing mutations.

Ultraviolet light This is the most common non-ionizing radiation, the energy of which is used in converting the bases into more reactive forms.

Chemical mutagens Most of the chemical carcinogens are mutagens which may be both organic and inorganic molecules. Some organic substances interact directly with DNA molecule and other organic and inorganic substances require a metabolic activation to become mutagenic. The inorganic compounds include arsenic, asbestos, cadmium, chromium, etc.

Biological mutagens Many of the pathogenic microorganisms like influenza virus are capable of producing antigenic mutation. These microbes undergo mutations very often. But respective antibodies are not available in man so that epidemics result in human populations.

MUTAGENICITY EVALUATION

There is a high correlation between mutagenic and carcinogenic properties of chemicals. Mutations are unable to cure so that an urgent need is warranted to detect mutagenic chemicals. Disciplines like genetics, biochemistry and cell biology have developed short-term tests to detect mutagens. More than 100 tests are currently available using a wide range of organisms but no single test can detect all mutagenic effects. Therefore, mutagenicity can be evaluated by employing a combination of different tests. The detection of mutagenic substances involves three criteria, namely, sensitivity, specificity and predictive value. The mutagenicity tests are grouped under the following four categories.

Detection of Primary Damage in DNA (Indirect Measurement of DNA Damage in the Cell) (Sister Chromatid Exchange Test)

The cells are allowed to grow for some time in thymidine analog namely, Bromodeoxyuridine (BrdU) which incorporates into the newly synthesized DNA instead of thymidine. In the cells which have passed through two cycles of DNA synthesis in BrdU, exchanges can be observed in the metaphase stage because the chromatids of one chromosome have incorporated different amounts of BrdU. This can be detected by using fluorescent chromium dye and scanned for chromosomes.

Gene Mutation Test (Ames' Test)

This is the best known mutagenicity test using certain bacterial strains of *Salmonella typhimurium*. This bacteria requires histidine for its normal growth in the culture media due to its specific mutation. When a mutagenic substance is added to the medium, DNA is mutated at the same locus so that they regain the capacity to produce histidine (back-mutations or reversions). These microbes are able to form visible colonies on histidine-deficient agar plates within a few days. The number of colonies per plate is a measure for the mutagenic potential of the test chemical.

Standard Bacterial Tester Strains

TA1535 It is a strain with histidine mutation and is used to detect mutagens causing base pair substitution.

TA1537 and *TA1538* These are also the strains with histidine mutation and are used to detect various kinds of frameshift mutagens.

TA100 and *TA98* These strains are much more sensitive to carcinogens and so are useful to detect carcinogens.

Three standard tester strains, namely, TA1535, TA1537 and TA1538 are used in combination with strains TA100 and TA98 in generalized mutagenesis tests.

Detection of Chromosomal Aberrations (Micronucleus Test)

Due to chromosomal breaks, the chromosomal fragments without centromeres are sometimes seen outside the interphase nucleus in one of the daughter cells. After the completion of cell cycle, the extranuclear DNA fragment forms a micronucleus. The number of micronuclei per thousand binucleated interphase cells indicate the induced genotoxic damage.

Detection of Potential for Proliferation

These tests measure the potential of chemical substances to induce proliferation in normal mammalian cells. However, these tests have less practical applications due to poor reproducibility.

REVIEW QUESTIONS

1. Define mutagens and mutagenicity.
2. Classify mutations and mutagenic agents.
3. Describe the major groups of mutagenicity tests.

16

SAFETY EVALUATION
OF TOXICANTS

Industrial chemicals, pesticides, food additives, pharmaceutical and veterinary drugs, etc. are the substances which bring adverse effects on humans directly or indirectly. As a result of tragedies and disasters, many regulations have been developed by the regulatory authorities. The analysis of toxicological results is useful to identify the hazardous nature of a chemical which can be tested through a number of toxicity studies. From the toxicological point of view, **hazard** is the biological property of the chemical in interacting with the species concerned. **Risk** is a statistical term which expresses the probabilities of the hazard. Hazard of the toxic substances can be determined by experiments but the risk cannot. The **no-observed effect level** (NOEL— the dose at and below which no toxicity occurs) is the only evidence of toxicity and is the first parameter to be used in risk assessment.

RISK MANAGEMENT AND MONITORING

Risk management involves the minimization of risks associated with the hazard and is accomplished in three ways. The first practice is risk avoidance in which the measures are taken in such a way that the risks associated with a particular hazard are not expressed. In the second phase, all measures are taken to ensure that the risks are kept as low as possible. Thirdly, various safety standards are adopted in practice. Risk monitoring involves periodical observations to check whether the recommended risk management steps are in practice or not.

ENVIRONMENTAL HAZARDS AND RISK ASSESSMENT

Environmental hazards are not restricted only to test organisms but also to other similar organisms with similar habitats. Therefore, test animals serve as indicator species as well as sensitive organisms for the overall trophic organization of the habitat. The effect of toxicants on the environment is studied by **ecotoxicity tests** by which the impact of toxicants on organisms, the fate of chemicals in the environment, and their distribution and effect on other chemicals are also known. These tests are also useful for evaluating the concentration of the toxicant, the time required to bring out the effect and the adverse effects of toxic substances on organisms. **Bioassay** is a toxicity test by which the potential of a toxicant to elicit the effect on organisms is evaluated. In other words, it is a test to determine the concentration of a chemical from the response exhibited by the test animals.

CRITERIA FOR SAFETY EVALUATION

The safety evaluation of a toxicant requires some criteria to express the toxicity which is to be quantitatively measured. In most acute toxicity tests, **mortality** is the criterion to evaluate the toxicity of a chemical. In these tests, groups of test organisms of the same size, weight and age are exposed to various concentrations of the toxicants for a specific period of time. The mortality percentage of the test animals in each concentration is plotted on a graph as shown in Figure 16.1.

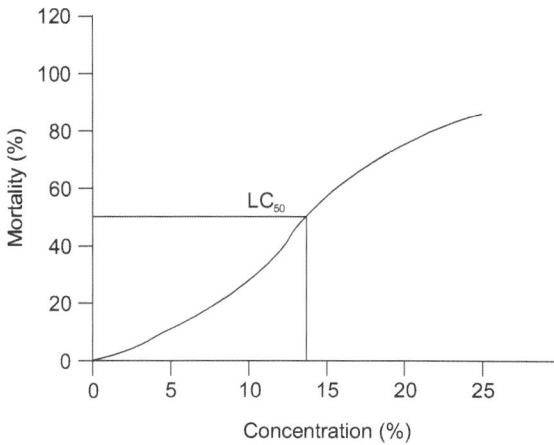

Figure 16.1 Mortality of test animals on exposure to various concentrations of a toxicant

In the graph, each point shows the response of the organism to a particular concentration and the least variability is found at 50% level of the response. Therefore, the concentration at which 50% of the test animals respond at a specific exposure period is an indicative of the toxicity of the substance (median lethal concentration).

Upper and Lower Confidence Limits

From the LC_{50} value, the upper and lower confidence limits can be calculated in the following manner.

The *f* value (*f* represents fidutial limit) is calculated by applying the following formula.

$$\log f_{95} = \frac{2.77}{\sqrt{N}} \times \log S$$

where,

$$S\,(\text{the arrived value}) = \frac{LC_{84}}{LC_{50}} + \frac{LC_{50}}{LC_{16}}$$

f_{95} = confidence limit (we can have 95% confidence in accepting the hypothesis) and

N = total number of animals tested at concentrations between LC_{16} and LC_{84}. (In a sigmoid curve, the mortality of the test animals approaches 0% as the concentration of the toxicant is decreased and approaches 100% as the concentration is increased but the region between 16% and 84% in the curve is linear).

Then the LC_{50} value is either multiplied or divided by the f_{95} value to obtain upper and lower confidence limits respectively.

$$\text{Upper limit} = LC_{50} \times f_{95} \text{ and}$$

$$\text{Lower limit} = \frac{LC_{50}}{f_{95}}$$

Slope of Safety

From the mortality curves for two different chemicals, it is clear that the LC_{50} values of chemicals x and y are the same but the slopes are different (Figure 16.2). The chemical x exhibits a flat slope and higher confidence limits whereas the chemical y exhibits a steep slope with lesser confidence limits. In both cases, the mortality of the test animals increases as the concentration of the chemical is elevated.

Thus the slope is an index showing the range of sensitivity to the chemical as well as the margin of safety.

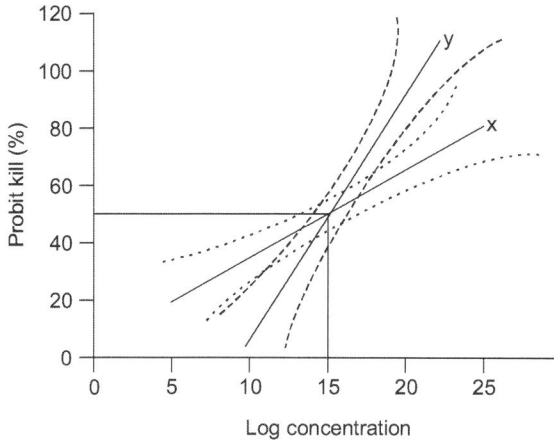

Figure 16.2 Mortality curve for two chemicals with upper and lower confidence limits

Cumulative Toxicity

The duration of the toxicity of a substance depends on the rate of its metabolic transformation and elimination. If these processes are not complete and if a second dose is given to the test animals, then the toxic effects of the first dose will be added to the second one, thus producing cumulative toxicity and the response produced is called **cumulative response.** To evaluate the cumulative toxicity, 100-day LD_{50} or LC_{50} index value is used.

$$100\text{-day}\,LD_{50}\,\text{index value} = \frac{LD_{50}(100\text{-day})}{LD_{50}(1\text{-day})} \times 100$$

This value indicates the percentage of dose which has to be given daily to the test animals to produce 50% mortality for a specific period. In other words, a higher index value indicates a greater percentage of dose which has to be given to produce 50% mortality over 100 days.

Table 16.1   CTF value, rating and nature of toxicants

CTF value	Rating	Nature
< 1.5	1	Practically non-hazardous
1.5–5	2	Slightly hazardous
5–10	3	Moderately hazardous
10–50	4	Very hazardous
50–100	5	Extremely hazardous
> 100	6	High extreme hazardous

The cumulative toxicity factor (CTF) is calculated by dividing the acute LC_{50} value by CLC_{50} or CLD_{50} value at weekly time intervals

$$CTF = \frac{Acute\, LC_{50}}{Cumulative\, LC_{50}}$$

The CTF value varies from 2.06 to 258.6 for various toxicants to which a numerical rating between 1 and 6 has been proposed as given in Table 16.1.

Calculation of Safe Level

The safe level of a toxicant can be calculated by using the formula given below.

$$Safe\, level\, of\, toxicant = \frac{LC_{50}\, 96\, hr \times 0.3}{S^2}$$

where,

0.3 = a constant and

$$S = \frac{LC_{50} \ 48 \ hr}{LC_{50} \ 72 \ hr}$$

EVALUATION OF COMBINED TOXICITY OF TOXICANT MIXTURES

The combined toxicity of pollutant mixtures can be evaluated by calculating CI values, CTC values, index 'V' values, MTI values and linear 'S' index values as described in the text.

The above types of toxicity evaluation studies will be in general highly useful in designing the safe level or tolerable level of pollutants to the aquatic biosphere. This in turn, could pave the way to establish limits and levels of acceptability of pollutants by the biotic components. Moreover, these tests are valuable to estimate the overall toxicity to screen test solutions for which toxicity data does not exist for selected test organisms. Therefore, the ultimate aim of toxicity evaluation is to predict the acceptable levels of toxicants in the environment to the biota.

ROLE OF LC_{50} OR LD_{50} VALUES IN SAFETY EVALUATION OF TOXICANTS

LC_{50} or LD_{50} value is a statistically determined parameter which indicates the lethal effect of a substance on test animals under experimental conditions and is used to compare the toxicity of substances with each other and to classify them according to their potential hazards. Though this value, gives information on the toxicity of a substance, it is mainly concerned with only one aspect namely, lethality. The toxicity of a substance should not be evaluated exclusively on the basis of this value, because the extrapolation of the value to a particular population of an animal belonging to different species will give entirely uncertain results. Moreover, the toxicity of pollutants is modified by various environmental factors because of which the extrapolation of the laboratory data to field conditions is rendered difficult and it should be undertaken with a great caution.

REVIEW QUESTIONS

1. Define and differentiate between 'hazard' and 'risk'.

2. Discuss the role of ecotoxicity tests in safety evaluation of toxicants.

3. Write down the procedure to calculate upper and lower confidence limits of a pollutant.

4. How are cumulative toxicity and safe level of toxicants calculated?

5. Describe the various parameters used to evaluate the combined toxicity of a mixture of pollutants.

17

CELL INJURY

As the cellular and subcellular changes form the basis for all disorders, the toxic effect of xenobiotics mainly rests on the cellular responses. The injury caused to cells by foreign compounds may be direct or indirect. Direct cell injury occurs when a toxicant interacts with one or more cell components. In the case of direct cell injury, the effect is due to a disturbance in the micro-environment of the cell. For example, the tissues have an insufficient supply of oxygen during hypoxia or anoxia so that the energy metabolism is disturbed leading to cell injury. The direct interaction of toxic substances with cellular metabolism may also lead to cell injury. The increased glycolytic activity during oxygen deficiency cannot meet the energy requirements of the cell. As a result, the energy-requiring processes such as protein synthesis, phospholipid metabolism and membrane transport processes are inhibited. On the other hand, the effect of toxicants on cellular metabolism via the manifestation of alterations in histology, physiology, biochemistry and activity levels of many enzymes constitutes indirect cell injury.

CHANGES DURING CELL INJURY

Cellular Oedema

A common morphological change during cell injury is the disturbed cellular water balance which is mainly determined by Na^+ and K^+ ions in the extra- and intracellular fluids. Reduced energy production will lead to decreased activity of these ions. Many toxicants cause membrane damage or disturbed energy metabolism of cells resulting in an increased uptake of water in the cells. This leads to the swelling of the cell or **cellular oedema** which is characterized by an increase in the cell size and by the foamy appearance of the cytoplasm due to swelling of intracellular structures. Cellular oedema is reversible, but becomes irreversible when it is not controlled in time. This will produce an increased cellular swelling, with cytoplasm containing numerous small or large vacuoles.

Hyperplasia

It is a condition in which there is an increase in the number of cells resulting in increased size and weight of a tissue or organ such as epithelial tissues (in kidneys, lungs and intestine), lymphocytes, bone tissues, etc. Epidermal hyperplasia causes thickening of various cell layers of the skin especially the keratin layer, i.e, increased number of cells at tissue or organ level.

Hypertrophy

It is also characterized by an increase in size and weight of a tissue or organ but in individual cells, i.e., increased size and weight at cellular level.

Atrophy

This indicates a reduction in size and weight of a tissue or organ due to a decreased amount of cytoplasm. Atrophy may be simple in which the decreased size and weight is due to the shrinkage of individual cells or a number of cells.

Neoplasia

It is the uncontrolled proliferation of cells resulting in the formation of a tumour, with morphological as well as functional changes in tissues or organs.

Necrosis

It is the breakdown of cell components after cell death, involving chemical processes namely, **autolysis** and **heterolysis.** Autolysis is the breakdown of macromolecules by lysosomal enzymes. Heterolysis is the accelerated degradation of the tissue by the lysosomal enzymes. Necrosis is of two main types, (i) coagulative necrosis and (ii) liquefactive necrosis.

Coagulative necrosis involves rapid degeneration as a consequence of lowering of cellular pH and of activation of the enzymes DNase and RNase. The nucleus becomes fragmented by the enzymes or a complete disintegration of cells is caused due to high autolytic or heterolytic processes so that the outline of cells and tissue structures are not recognizable.

Liquefactive necrosis involves swelling and bursting of cells, thus releasing their intracellular contents which can damage the surrounding cells.

If the cellular disturbance exceeds the limit of cellular homeostasis, cell death occurs because of irreversible cellular degeneration. Cell death may be due to two distinct patterns of morphological changes. One pattern is **necrosis** (already discussed above) which occurs in groups of cells rather than in individual cells, with exudative inflammation. Another pattern is **apoptosis** which is characterized by condensation of the cells regulated by physiological stimuli, involving no inflammation. Therefore, it is otherwise called as "shrinkage necrosis" or "physiological cell death" which involves relatively large number of cells.

SUBCELLULAR TARGETS OF TOXICITY

Chemical substances such as biological toxins, synthetic chemicals like pesticides, and heavy metals may cause cellular degeneration. With reference to cell death, the toxicants include two important types, of which one group could affect the transport system of membranes and the other interferes with mitochondrial oxidative phosphorylation. Though different kinds of toxic substances affect various sites in the cell, the main subcellular targets are the cell membrane, mitochondria, endoplasmic reticulum and nucleus.

Cell Membrane

Many biological toxins cause direct injury to the cell membrane thereby affecting its selective permeability. The membrane is indirectly damaged through peroxidation of lipids and covalent binding of reactive metabolites to the macromolecules. In general, the toxic substances affect the ionic permeability of the cell membranes, leading to cellular degeneration.

Mitochondria

Degeneration of mitochondria is due to a secondary effect of cell injury. Certain toxicants cause direct and specific injury to the mitochondria altering the mitochondrial metabolism. As a result, the production of ATP is blocked with the resultant inhibition of energy-requiring processes in the cell. The injured mitochondria also exhibit ultrastructural changes such as rapid condensation with loss of ions and water, swelling, disorganization of cristae, deposition of calcium salts, and so on. Lack of energy, in the cell leads to cell death.

Endoplasmic Reticulum

The substances which directly affect DNA-dependent RNA polymerase in the nucleolus, cause consequent separation of ribosomes from the rough-surfaced endoplasmic reticulum. As a result, the enzymes and proteins for normal functioning of the cell are not produced. This in turn would lead to degenerative processes in the cell membrane, mitochondria and other cellular organelles leading to cell death.

Cellular Calcium Homeostasis

Calcium homeostasis is mainly energy-dependent in which the cell membrane plays an important role. The necrotic cells are found to contain large quantities of calcium. In other words, certain toxic substances like heavy metals, endotoxins, etc. cause an increased calcium influx into the cells. The increased amount of calcium in the cytosol increases membrane permeability leading to cellular swelling and changes in the cytoskeleton and cell sap. It also causes swelling of mitochondria and endoplasmic reticulum. Moreover, calcium accumulates in the mitochondria resulting in mitochondrial calcification. The calcium ions, in the cytosol, also activate phospholipases which in turn bring damage to the cell and mitochondrial membrane.

Naturally, adaptive responses play an important role at the event of altered cellular homeostasis in order to minimize the effect of injury. These are usually reversible and are accomplished through different levels of metabolic and functional activity. If these adaptive responses fail, then the cell will be irreversibly damaged leading to cell death.

REVIEW QUESTIONS

1. Characterize cell injury.
2. Enumerate various kinds of cell damage.
3. Write an essay on the main subcellular targets of toxicity.
4. Altered cellular homeostasis leads to cell death. How?

18

TOXICITY AT THE LEVEL OF ORGANS AND ORGAN SYSTEMS

The adverse toxic effects of xenobiotics are mainly due to the physico-chemical reactions at the site of contact between the toxic substances and the body of organisms. These effects are called as **injuries** or **intoxication**. But in systemic effects, the substances are first absorbed into the body and then transported to various organs where they affect the functions of the organs or organ systems resulting in various symptoms. Factors such as local accumulation, transport of metabolites, local differences in covalent binding, organ-specific physiological and biochemical characteristics and the enzymes of biotransformation in an organ play a prominent role in bringing toxicity in organ systems. The metabolic activation of some substances can take place selectively in certain organs. Some other substances, which are metabolized in several organs, can cause toxic impact in only one of the organs. The effects of toxic substance on various organs and organ systems have been well elucidated in experimental animals like mice, rats, rabbits and guinea pigs as well as in humans.

The toxicants are found to affect each and every organ and organ system. Table 18.1 provides information about the overall defects caused by the toxic substances in different organs and organ systems.

Table 18.1   The overall disorders caused by toxicants in organs and organ systems

Organs/Organ system	Disorders
Digestive system	Stomach and bowel disorders, nausea, vomiting, abdominal pain and diarrhoea.
Stomach	Damaged stomach mucosa, mucosal necrosis, haemorrhage, disintegration of gastric glands and inhibition of protein synthesis in gastric cells.
Intestine	Inhibition of cellular protein synthesis, intestinal fibrosis, alterations in the permeability of membranes, enzyme production, and damage in brush border.
Liver	Liver damage, hepatic cell injury, structural alterations and disorganization of hepatocytes, nuclear degeneration and necrosis, damaged lysosomes, enzyme synthesis and disruption of mitochondrial cristae in the hepatocytes.
Respiratory system	Altered respiratory process, wheezing, bronchitis, respiratory depression, pneumonia, and pulmonary fibrosis.
Nose	Sinus problem, bleeding, discomfort, irritation and necrosis in nasal mucosa.
Lungs	Congestion of lungs, lung carcinoma, lung abscesses and pathological lung functions.

(Contd.)

Table 18.1 [Continued]

Organ/Organ system	Disorders
Circulatory system	Low blood pressure, cardiac failure, anaemia, leukaemia, Hodgkin's disease, neoplasms and vascular proliferation.
Blood	Decreased blood cell counts and haemoglobin content and inhibition of immune system.
Heart	Heart damage and cardiac irregularities, degeneration of cardiac muscle, hypertrophy of ventricles and reduced synthesis of protein, RNA and DNA in the ventricles.
Blood vessels	Aortic calcification, increased capillary permeability, microvascular injury and peripheral vascular lesions.
Nervous system	Paralysis, neurological impairment, encephalopathy and neuropathy.
Brain	Inhibition of cholinesterase activity and shrinkage of brain cells.
Neurons	Alteration in Ca^{2+} ion concentration, neurofibrillar pathology of axons, axonopathy with demyelination, perivascular myelin swelling, pericellular oedema, plasmolysis and karyolysis of neurons and inhibition of cholinesterase activity.
Spinal cord	Lesions and changed reflex conditions.
Endocrine system	Impaired growth, decreased basal metabolism, reduced thyroid function, altered sugar metabolism due to disturbed pancreatic function and disturbances in renal function and fertility.

(Contd.)

Table 18.1   (Continued)

Organs/Organs system	Disorders
Pituitary gland	Pituitary adenoma.
Adrenal gland	Increased weight, induced plasma epinephrine, hyperglycemia, vacuolization, and hypertrophy and alterations in adrenal functions.
Thyroid gland	Disturbed thyroid function, hyperplasia, glandular atrophy, goitre and injury to thymic epithelial cells and thymocytes.
Excretory system	Kidney damage, kidney diseases, lipid nephrosis, abnormalities in glomerular epithelial cells and basement membrane, carcinoma, nephropathy, necrosis of convoluted tubules and tuberculoma.
Reproductive system	Fertility impairment.
Male organs	Atrophy and necrosis of seminiferous tubules, necrosis of testis, vacuolization of germinal cells, lack of differentiation and maturation of spermatocytes and reduced secretion and maturation of spermatozoa in cauda epididymis.
Female organs	Reduced protein level in the uterus, reduced activity level of enzymes, decline in ovarian DNA and RNA, altered protein and nucleic acid metabolism and teratogenic effects.

DERMATOTOXICITY

The skin is one of the major organs for the entry of toxic substances into the body of animals. Three distinct layers can be recognized in the skin, namely, epidermis (a keratinized squamous epithelium), dermis (a thick underlying fibro-elastic connective tissue) and hypodermis (a subcutaneous layer of loose connective tissue containing adipose tissue). The epidermis is the only layer in the skin, which permits the penetration of toxicants into the body. The entry of the substances through the skin depends on species, localization, temperature, pH and health of the individuals as well as the ionic state, molecular size, lipophilicity, viscosity and concentration of substances. The skin provides the largest surface area and forms an important target organ for many toxic substances. The toxicants are found to cause two major kinds of toxic effects in the skin, namely, irritation (direct injury) and allergy (immunological reaction). These impacts are mainly due to local exposure involving only limited number of reactions such as inflammation, degeneration and proliferation or some combinations of these processes. When a toxicant comes in contact with the skin, it may react with the cells of the skin causing skin irritation followed by dermatitis, or may invade the skin and react with local proteins or may penetrate into the skin to be distributed causing systemic effects.

Dermatotoxic effects can be studied under the following headings:

Primary Irritation (Irritant dermatitis)

The irritant dermatitis is the most common toxic skin reaction due to the exposure to detergents, soaps, many organic substances, etc. This includes mild, severe and moderate contact dermatitis leading to inflammation of the skin and causes pain. Table 18.2 gives an account of various types of irritant dermatitis and their extent of damages.

Table 18.2   Irritant dermatitis and characteristic damages

Types	Damage	Characteristics
Mild contact dermatitis	Hyperkeratosis	Proliferation of stratum corneum and thickening of skin.
	Hypergranulosis	Proliferation of stratum granulosum.
	Acanthosis	Proliferation of stratum spinosum.
Moderate contact dermatitis	Eczema	Combined effect of degeneration, proliferation and inflammation, vasodilation, oedema, exudation and migration of leucocytes, thickening of skin, and formation of blisters and abscesses in the epidermis.
Severe contact dermatitis (corrosion)	Corrosive lesions	Inflammatory reactions with cell death in epidermis and dermis, proliferation of epithelial and connective tissue cells, and shedding of dead cell layer with a scar at the site.

Allergic Dermatitis

This involves the delayed-type allergic reaction in which the substance, usually a hapten, makes its first entry into the horny layer of the skin, binds with certain proteins and becomes a complete **allergen**. This activated substance now binds to T lymphocytes and is transported to the local lymph gland. This process is termed as **sensitization**. At the event of the subsequent entry of the substance, the sensitized T lymphocytes produce lymphokines, which cause inflammatory reaction at the site of contact or anywhere in the skin. This results in vasodilation, perivascular accumulation of leucocytes, hyperplasia, and may persist for years without any symptoms. The substances that cause allergic dermatitis include dichromate salts, epoxy resins, hexachlorophene, formaldehyde, metals like nickel, chromium and cobalt, aromatic amines, ethylene diamide, local anaesthetics, cement, cosmetics, etc.

Phototoxic and Photoallergic Reactions

The disorders caused by the combined effect of the toxic substance and sunlight are referred to as **phototoxic reactions** or **phototoxicity**. Based on the site of the toxic effect, it is distinguished into local phototoxicity and systemic phototoxicity. The local phototoxic reactions occur at the site where the skin comes in direct contact with the toxic substances like soot, tar products, substances of vegetable origin and certain dyes (eosine, methylene blue, etc.). In systemic phototoxicity, the toxicants enter into the body through the routes other than the skin, then reach the skin via the blood and cause reactions in the skin on exposure to light. Drugs like tetracycline, sulphonamides, porpyrines, etc. are found to cause phototoxic effects. In the case of photoallergic reaction, the binding of a hapten with skin protein is catalysed by the photochemical reaction so that the substance produces a reaction at the site on the skin when exposed to sunlight. The causatives of photoallergic reactions include phenothiazine derivatives, sulphonamides, quinine, thiocarbamine, cyclamates, halogenated phenols, etc.

Skin Tumours

The skin tumours may be epithelial or mesenchymal in nature and are caused by soot, tar, creosote oil, cutting oil, arsenic and their compounds, polycyclic aromatic hydrocarbons, heterocyclic compounds, UV light, ionizing radiations, and so on. The types, nature and characteristics of skin tumours are given in Table 18.3.

Table 18.3   Various types of skin tumours

Types	Nature	Characteristics
Papilloma	Benign	Epithelial tumour in skin, and cutaneous mucosa and ulceration.
Squamous cell carcinoma	Malignant	Epithelial tumour in squamous epithelium, external damage with haemorrhage and necrosis, slow metastasis.
Basiloma	Benign	Epithelial tumour in the stratum basale.
Epithelioma	Benign	Keratinizing epithelioma, or necrotic- and calcifying epithelioma or tricho-epithelioma.
Fibromas	Benign	Mesenchymal tumour in the connective tissue cells; lobed or solid, or solitary or multiple with variable sizes.
Fibrosarcoma	Malignant	Mesenchymal tumour in the connective tissue cells; infiltration into the surrounding tissues.
Melanoma	Malignant	Mesenchymal tumour in melanocytes, uneven distribution, more fatal.

A number of toxicants such as heavy metals like silver, bismuth, arsenic and mercury, drugs like phenothiazines, acridines and 4-aminoquinoline, and alkylating agents like bisulphate are shown to exert melanotoxic effect on the skin. Some substances cause local dispigmentation of the skin (hypopigmentation) and others cause local or diffuse hyperpigmentation. The lesion in the hair follicle is called acne, which causes inflammatory reaction in the dermis due to exposure to some oils and pesticides like polyhalogenated biphenyls. Substances like trafuryl, cobalt chloride, etc. cause **urtricaria** in the skin resulting in itching, pain, erosions, oedema, lichenification and eczema.

GASTROINTESTINAL TOXICITY

Toxicity may produce various proliferative changes in the gastrointestinal tract as it is the main penetration route for the toxic substances. The intoxication of substances on the gastrointestinal tract may result in direct injury to the mucosal cells, or in the interaction with receptors, or in reducing the peristaltic movement or in carcinogenesis. The major injuries caused by the toxic substances in different regions of the gastrointestinal tract are summarized in Table 18.4.

Table 18.4 Major injuries in the gastrointestinal tract caused by toxicants

Region	Injuries
Oesophagus	Irritation, inflammation, epithelial necrosis, thickening of epithelium, reflux, and chemical oesophagitis, papillomatous hyperplasia, neoplasia, papillomas and squamous cell carcinoma.
Stomach	Erosion and ulceration of stomach wall, diffuse gastritis, dysplasia, gastric carcinoma, adenomas and adenocarcinomas.

(Contd.)

Table 18.4 (Continued)

Region	Injuries
Small ntestine	Excessive loss of fluid or electrolytes, malabsorption, vomiting, diarrhoea, erosions and ulcers in mucosa, enteritis, atrophy of villi, adenocarcinoma, lymphosarcoma and intestinal metaplasia.
Large intestine	Similar to that found in stomach and small intestine, ulcerative colitis and colo-rectal adenoma.

Table 18.5 gives an account of different categories of toxicants that cause harmful effects on the gastrointestinal tract.

Table 18.5 Various categories of gastrointestinal toxicants

Types of toxicants	Examples
Food and food additives	Glucosinolates, nitrites, nitrates, nitrosamines, pesticide residues, saccharine, sorbitol, butylated hydroxyanisole, etc.
Drugs	Phenyl butazone, antibiotics like tetracycline and penicillin, metronidazole, anti-inflammatory drugs, beta-blockers, digitalis glycosides, analgesics, corticosteroids, etc.
Environmental and industrial toxicants	Fluorides, corrosives, detergents, halogenated aliphatic and aromatic hydrocarbons, aldehydes, alkylating agents, aromatic amines, etc.

(Contd.)

Table 18.5   [Continued]

Types of toxicants	Examples
Metals	Calcium, lead, aluminium, cadmium, sodium, zinc, iron, magnesium, barium, chromium, mercury, etc.
Carcinogens	Nitrosamines, nitrates, polycyclic aromatic hydrocarbons, polychlorinated biphenyls, dimethyl hydrazine, sulphate compounds like dextran sulphate, etc.

HEPATOTOXICITY

In toxicological point of view, liver plays an important role because all substances absorbed by the gastrointestinal tract pass through it before entering into the general circulation. It is involved in metabolism, detoxication, secretion and excretion, so it is highly vulnerable to toxic substances. Hepatotoxins may exert direct effects in the form of primary lesions and disturbed cell metabolism or indirect effects in which they interfere with cell metabolism resulting in the loss of cell integrity. The acute cytotoxic effects include steatosis (fatty degeneration of liver), parenchymatous degeneration, hydropic degeneration, apoptosis, and so on. Chronic active hepatitis, subacute liver necrosis, steatosis, phospholipidosis, fibrosis, cirrhosis (micronodular or macronodular), etc. are chronic disorders caused by the hepatotoxins. In addition, some substances cause hepatotoxicity through the immune system leading to delayed immunological reactions. Table 18.6 shows the major groups of chemicals that induce hepatotoxicity in test animals as well as in man.

Table 18.6　Hepatotoxic chemicals and their effects

Hepatotoxic chemicals	Effects
Analgesic and antipyretic drugs (acetaminophen)	Chronic hepatitis.
Anti-inflammatory drugs (phenylbutazone)	Acute hepatitis, choleostasis and hypersensitivity reaction.
Anaesthetics (halothane)	Centrilobular necrosis and immune-dependent hepatotoxicity.
Antibiotics (tetracycline) and others like methotrexate, methyldopa, isoniazid, etc.	Acute and severe hepatitis, steatosis, choleostasis, cirrhosis and liver necrosis.
Alcohol	Liver damage, cellular necrosis, steatosis, cirrhosis, acute hepatitis and collagen formation.
Carcinogens Natural—aflatoxin B, safrole, cycasin, etc. synthetic— dimethylnitrosamine, DDT, PCBs, CCL4, chloroform, vinyl chloride, etc.	Adenomas and carcinomas.

RESPIRATORY TRACT TOXICITY (INHALATION TOXICITY)

In higher animals, the respiratory tract constitutes complex system, the important function of which is the uptake of oxygen and elimination of carbon dioxide. It consists of air-conducting regions, namely, nose, pharynx, larynx, trachea, bronchi and bronchioles, and a respiratory region, the alveoli. The air-conducting regions transport air into the system whereas the respiratory region is responsible for gaseous exchange. The respiratory tract provides a large surface area and a thin barrier between air and blood so that the lungs play a role in the uptake of gases, vapours and volatile as well as non-volatile substances.

Absorption of Gases and Vapours

In general, gases and vapours are absorbed through the entire respiratory system and their site of deposition depends on the solubility especially liposolubility or the reactivity of the gas. When a gas with high water solubility passes through the respiratory tract, a considerable amount of the gas is absorbed by the mucous and tissue, and a part of the absorbed gas is released through the exhaled air. If the gas has low chemical reactivity, then a portion of the gas enters the bloodstream via the tissue and distributed throughout the body. If it possesses a high chemical reactivity, the gas is converted into reaction products in the mucous or epithelial layer and removed through the mucous or exhaled air.

The low solubility gases are absorbed depending on their chemical reactivity so that a gas with low chemical reactivity will pass through the alveolar epithelium with the resultant equilibrium between the blood and other tissues. In this case, toxic effects are exerted both on the walls of the respiratory tract as well as on the other tissues. On the other hand, the toxic effect will be restricted only to the respiratory system, if the gas has high chemical reactivity. The gases having intermediate solubility and low chemical reactivity exhibit marked exchange with the blood and so the body has high absorption capacity to these gases. In this way, the gases and vapours that possess low chemical reactivity do not accumulate, whereas chemically reactive gases and vapours and their metabolites are toxic, and accumulate in the tissues, if they are non-volatile.

Deposition of Aerosols

Aerosols are the stable mixtures of air and solid dust particles or droplets of fluid. If the aerosols contain particles of the same size, then they are called as **monodisperse aerosols** and those, which possess particles of varying size, are referred to as **polydisperse aerosols**. The polydisperse aerosols are used in most of the inhalation toxicity studies as man is always exposed to these aerosols. These particles are transported into the respiratory tract mainly by convection and so their deposition on the system mainly depends on the strength of the air flow and the particle size. They are deposited through several processes such as impaction, sedimentation, diffusion and interception. Deposition of the particles occur by impaction in nose and larynx, by sedimentation in the trachea, bronchii and bronchioles, by diffusion in the alveoli, and by interception in the walls of the respiratory tract. The site of deposition of the inhaled particles depend on their size and shape. For example, particles with the size of 5–30 μm are deposited in the nasopharynx, particles of 1–5 μm in the trachea and bronchii and particles with the size smaller than 1 μm in the lungs. Fibriform particles such as asbestos are easily deposited in the lungs as they lack adequate clearance mechanism.

Pathological Effects in the Respiratory Tract Caused by the Toxic Substances

The reactions of the inhaled gases and particles with the respiratory tract can be categorized into direct toxic effects or pharmacological reactions and indirect effects through immune reactions. These reactions mainly depend on the chemical properties of the substances, their solubility in water,

Table 18.7 Direct effects of toxic substances in the air passage

Regions of air passage	Pathological effects
Nose	Degeneration, proliferation, inflammation of nasal and olfactory epithelium, squamous metaplasia, goblet cell hyperplasia, nasal cancer, squamous cell carcinoma and adenocarcinomas.
Lower air passage	Squamous metaplasia, hyperplasia, fibrosis, suppression of mucociliary clearance, asthma (bronchoconstriction) and bronchial carcinoma.
Trachea and bronchi	Alveolitis—desquamation and proliferation of pneumocytes, metaplasia and formation of hyaline membranes.
Alveoli	Alveolar bronchiolization—pulmonary tumour such as bronchoalveolar adenomas and carcinomas.
	Emphysema—vacuolization due to damaged alveolar walls and fusion of alveoli and oedema.
	Lipidosis—foamy cells in the alveolar spaces.
Lungs	Pneumoconiosis—accumulation of inorganic particles in the lungs and anthracosis.
	Silicosis—granulomas and fibrosis.
	Asbestosis—thickened pleura and fibrosis.
	Lung cancers—squamous-cell carcinoma, small-cell bronchial carcinoma, adenocarcinoma and large-cell bronchial carcinoma.

size of the particles and the sensitivity of the respiratory epithelium. However, both the types of effects culminate in the degeneration, proliferation and inflammation in the respiratory tract. But as there are differences in the effects of inhaled gases or particles between humans and experimental animals, the results obtained from the animal experiments cannot be manipulated to man.

The inhaled particles are removed from the respiratory tract by two important mechanisms, namely, **mucociliary clearance** and **alveolar clearance**. Abnormal functioning of these mechanisms would lead to the accumulation of the particles in the lungs resulting in reduced respiratory capacity, toxic effects, and infections. The clinical symptoms of the respiratory tract damage include altered respiratory movements, noise during respiration due to the constriction of airways or due to the obstruction by inflammatory exudates, cyanosis (blue colouration of skin and mucous membrane), heaviness of chest, and so on. Table 18.7 gives an account of various pathological alterations due to the direct effect of toxicants in the air passage.

Indirect Effects Through Immune System

The reactions of immune system to remove toxic, volatile and particulate substances from the body are inadequate so that these substances cause highly undesirable effects in man through excessive reactions and tissue damage. Such unwanted reactions are called **allergic** or **hypersensitivity reactions** which are of four major types. Of them, Type I hypersensitivity results in local allergic reactions such as hay fever, asthma, etc. due to exposure to drugs, dust particles, pollen, formalin, flower and plant oils, and so on. Type II hypersensitivity is not concerned with inhalation toxicity. Type III hypersensitivity results in the formation of immunoglobulin complexes (IgM and IgG) on exposure to dust particles or in an immune disease. Instead of being cleared, these complexes are deposited in the kidneys, arteries, joints, skin, and lungs and cause inflammation, fibrosis, and destruction of organs. Type IV hypersensitivity or delayed-type hypersensitivity is due to chronic infectious diseases caused by bacteria, fungi, protozoa and endoparasites and is not concerned with inhalation toxicity. However, it is shown that the contact hypersensitivity to certain metals result in granuloma of lungs in tuberculosis.

Tests for Respiratory Function

The tests for respiratory function are useful to detect and quantify the pathological alterations caused by the toxic substances in the respiratory tract. The tests for inhalation toxicity, in general, include the routine toxicity tests such as acute, subacute, subchronic and chronic toxicity tests. In addition to these standardized methods, the substances can also be studied for their more specific harmful effects such as changes in the properties of the system and effectiveness of gas exchange. Studies on alveolar macrophages, bronchoalveolar lavage fluid, lung homogenate, connective tissue of lungs, etc. as well as on teratogenicity, carcinogenicity and mutagenicity are made to evaluate the parameters like rate of breathing, tidal volume of air, total capacity of lungs, residual volume of air in the lungs, blood-gas analysis, etc.

Exposure or inhalation chambers In the tests for respiratory function, all the experimental animals must be exposed to inhale the same concentration of the substance. Therefore, the substance under investigation would be equally distributed in experimental conditions. In order to achieve this protocol, **exposure** or **inhalation chambers** are used and they help to expose the test organisms to a test atmosphere under controlled conditions. The test atmosphere may be an aerosol, or a gas, or a vapour, or their combinations. The chambers are made up of stainless steel, or aluminium, or plastics, or glass, and the design and construction of the chambers vary. The set-up of an exposure chamber and its accessories are depicted schematically in Figure 18.1.

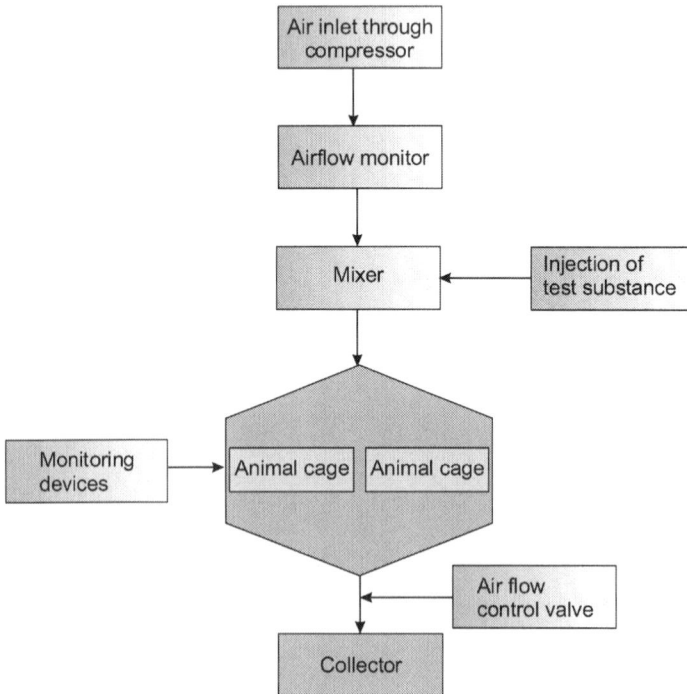

Figure 18.1   Schematic representation of an exposure chamber and its accessories

Monitoring devices   The concentration of aerosols can be measured by using the extracted test atmosphere from the test chamber through **gravimetric method** (weighing of filters before and after sampling) or **chemical methods**

(HPLC or colorimetry or gas chromatography). Special instruments such as impactors, substance meters, spray meters, etc. are also employed in measuring the concentration of aerosols. The particle size distribution of aerosols is determined with the help of special equipments and methods as given below.

Cascade impaction   Here, an equal amount of air is allowed to pass through narrow openings in different velocities with increasing steps. A small plate under each opening deflects the air flow so that the larger particles at high velocity precipitate first on the plate followed by smaller particles. Then the amount of aerosols in each plate is measured.

Aerodynamic particle size meter   In this equipment, the accelerated air is passed through a narrow opening and the difference in time taken for the passage of larger and smaller particles between two points is measured.

Condensation nucleus counter   In this instrument, the particles separated by electric field are passed along with a gas through a condenser and the gas-condensed particles are counted through optical systems.

Optical detectors   These devices measure the intensity of the dispersed light when the particles are passed through a beam of light and this represents a measure of the size of the particles.

Spray meters   When the particles pass through a beam of light, they obstruct the passage of the light and the deviation of the light is measured by spray meters.

CARDIOTOXICITY

The heart is an ideal pumping organ and possesses a number of intrinsic properties but is not the usual target

Table 18.8   Cardiotoxic substances and their effects

Cardiotoxicants	Effects
Halogenated hydrocarbons like volatile anaesthetics	Lowering of heart rate (pulse), contractility and conduction, cardiac depression and cardiac arrest.
Ethanol	Reduced contraction, ventricular fibrillation, sudden death.
Heavy metals (cadmium, cobalt and lead)	Cardiomyopathy and degenerative alterations.
Digitalis glycosides	Prolonged refractory period in the AV node, decreased rate of impulse conduction, re-excitation in the SA node, automaticity, arrhythmia, ventricular fibrillation and cardiac arrest.
Epinephrin and its analogues	Decreased contractility and endocardial necrosis.
Tricyclic antidepressants (imipramine, amitriptylene)	Ventricular arrhythmia.
Neuroleptic substances (phenothiazone, butyrophenone derivatives)	Tachyarrhythmias.
Psychoactive substances (amphetamine, cocaine)	Increased heart beat and blood pressure.
Antibiotics like anthracyclines	Arrhythmia and congestive cardiomyopathy.

organ for the toxic substances. However, a number of xenobiotics affects its anatomy and physiology resulting in abnormal heart functioning through alterations in the rate of heartbeat, conduction, excitability or conductivity. Chemically-induced morphological changes include hypertrophy (increased weight of the heart), cardiomyopathy (failure of functioning of heart muscle) and myocardial necrosis. Two major functional abnormalities caused by the toxic substances are arrhythmias and contractility disturbance. In arrhythimia, the heart beat is very slow or very fast due to altered impulse transmission, resulting in extrasystoles, ventricular fibrillation or total cardiac arrest. Contractility disturbances that are due to degenerative alterations with lowering of contractibility in the heart muscle leading to cardiac depression and arrest. The biochemical changes include increased activity levels of certain enzymes such as lactic dehydrogenase, aspartic aminotransferase and creatine phosphokinase. All the cardiotoxic substances are found to be arrhythimeogenic and are given in Table 18.8 with their effects.

HAEMATOTOXICITY

The bone marrow is highly sensitive to the toxic chemicals. The direct effect of toxicants on bone marrow could result in dysplasia (abnormal production of cells), hypoplasia (decreased production of cells), aplasia (inhibition of cell production and malignancy of the cells). The inhibition of the red cell line causes severe anaemia whereas the inhibition of the white cell line results in the reduction of granulocytes leading to increasing susceptibility to infections. The number of blood cells are affected directly or indirectly by the toxic substances that may cause haemolysis either in the bloodstream or in the reticuloendothelial system resulting in disturbed haemoglobin production, abnormal membrane structure and deficient glycolytic enzymes. Some substances reduce the oxygen-carrying capacity of the RBCs resulting in methaemoglobin. The haemotoxic agents include a number of drugs and chemicals such as antibiotics (chloramphenicol), anti-rheumatic agents (phenylbetazone), chlorpromazine, thiouracil, sulphonamides, etc.

Malignant blood disorders are due to chronic exposure to certain chemicals like benzene, asbestos, heavy metals, pentachlorophenol, chlorophenol, polystrene, vinyl chloride, ionizing radiation, etc. These diseases fall under two categories, namely, leukaemias and lymphomas. Leukaemias cause abnormal production of cells in the bone marrow and include myeloid leukaemias (acute myeloid leukaemia, acute monocytic leukaemia, chronic myeloid leukaemia and erythroleukaemia) and lymphatic luekaemias (acute and chronic). Lymphomas are malignant diseases with abnormal cells in the lymph nodes and include Hodgkin's disease (giant cells in the lymph nodes and other organs), multiple myelomas (proliferation of plasma cells, especially in bones) and non-Hodgkin's lymphomas (proliferation of lymphatic cells). Toxic compounds like benzene as well as ionizing radiation cause injury to chromosomes resulting in aneuploidy, chromosomal aberrations, abnormalities in cell cycles, and so on. All these alterations may lead to haematological malignancies, especially leukaemias. The following Table 18.9 indicates major haematotoxins and their ill effects on man.

Table 18.9 Major haematotoxins and their effects

Haematotoxins	Effects
Antihypersensitive agents (methyldopa)	Autoimmune haemolytic anaemia.
Antibiotics (penicillin)	Immune haemolysis.
Drugs (quinine, quinidine)	Intravascular cell destruction through immune complex mechanism.
Aromatic compounds containing amino-, nitro- or hydroxy groups.	Denaturation and precipitation of Hb (Heinz bodies)
Nitrates, chlorates, quinines, aniline or sulphonamide	Methaemoglobinaemia
Oxidants like sulphonamides, phenacetin or acetamilide	Sulphaemoglobinaemia
Carbon monoxide	Decreased oxygen-carrying capacity of Hb.
Cyanide	Hyperpnea (rapid breathing), arrhythmia, respiratory failure and death.
Benzene	Anaemia, thrombocytopenia, granulocytopenia and leukaemia.
Lead	Anaemia, reduced lifespan of RBCs and reduction in number of stem cells and plasma cells.
Iron	Oxidative haemolysis.

NEUROTOXICITY

The nervous system is anatomically divided into the central nervous system (brain and spinal cord comprising neurons and glial cells) and peripheral nervous system (afferent and efferent nerve fibres). It functionally includes somatic nervous system and autonomic nervous system. The somatosensory system contains afferent nerve fibres and the somatomotor system possesses the efferent nerve fibres. The sensory neurons conduct the stimuli from the sense organs to the central nervous system and the motor neurons transmit the impulses from central nervous system to the effector organs. The transmission of nerve impulses at the neuromuscular junctions and synaptic regions occurs via the neurotransmitters. Each neuron or nerve cell consists of a cell body (the soma) and two types of processes namely, axons and dendrites. The axon conducts the nerve impulses from the cell body to the other neurons or to the effector organs via dendrites. Long chains of protein molecules (neurofibrils) are found throughout the length of the axons and dendrites. These molecules play a major role in the axonal transport, the disruption of which would cause impairment in the functional integrity of the neuron.

The nervous system is the most sensitive one for many toxic chemicals which interrupt the electrophysiological mechanisms of the neuron. Toxicants also cause structural changes in the axon, myelin sheath and cell body. The neurotoxicity depends on the dose and route of exposure of substances as well as the species. The neurotoxic effects can be classified based on the site of interaction of toxicants in the body of the organisms. The lesions caused by a substance in the nerve cell body is called as **neuronopathy**, in the axon as **axonopathy**, in the synapse as **synaptopathy** and in the myelin sheath as **myelinopathy**. Table 18.10 gives an account of selective toxic impacts caused by a number of neurotoxic substances.

Table 18.10   Selective neurotoxicity caused by neurotoxins

Neurotoxins	Selective neurotoxic effects
Methyl mercury	Loss of ribosomes and disintegration of Nissl bodies, destruction of entire neuron, Minamata disease—lesions in central nervous system, peripheral nervous system and cerebral cortex, disruption of visual and sensory function, ataxia.
Cyanide and azide	Cytotoxic anoxia.
Aluminium	Degradation of neurofibrils.
Insecticide	Blockage of conduction of nerve impulses, paralysis, sensory impairment, epileptic symptoms due to impairment of central nervous system.
Barbiturates	Anoxia in the brain especially in cerebral cortex.
Glutamate and kainic acid	Disturbed ionic homeostasis in the membrane.
Adriamycin	Inhibition of RNA and protein synthesis in neuron.
Vinca alkaloid	Accumulation of neurofibrils in axons, blockage of axoplasmic transport.
Carbon monoxide	Sclerosis of white nerve tissue, brain damage.
Lubricating oil	Polyneuropathy.

Axonopathy

An axonal lesion which occurs very near to the cell body is called proximal axonopathy and those which occur close to the nerve endings form distal axonopathy. Chemicals like β-iminodiproprionitrile (IDPN) cause proximal axonopathy resulting in further slowing down of delayed axonal transport to the proximal axon resulting in swelling. The distal axonopathy causes direct damage to the neurofilaments especially the distal parts of long axons and delayed axonal transport of enzymes and other proteins to the neurofilaments from the soma via the axoplasm. Chemicals such as acrylamide, carbon disulphide, *n*-hexane methyl *n*-butyl ketone and organophosphates bring out degeneration in the distal axons. The distal axonopathy due to certain chemicals like tri-*ortho*-cresyl phosphate exhibit accumulation of smooth endoplasmic reticulum prior to the degeneration of distal axon whereas the chemicals such as carbon disulphide, acrylamide and *n*-hexane produce axonal swelling due to the accumulation of the neurofilaments.

Myelinopathy

It refers to the loss of the myelin sheath (demyelination) due to the direct toxic action of substances on the membrane structure. It is also caused by the inhibition of enzymes involved in water and ionic transport and oxidative phosphorylation with the resultant reduction in the level of ATP and chelation of metals. Chemicals such as triethyltin, hexachlorophene, etc. exert direct effect on the myelin sheath resulting in fluid-filled vacuoles between the myelinic lamellae. Certain neurotoxins like lead and triparanol act on oligo-dendrocytes which surround the myelin sheath as well as Schwann cells. The demyelination can be easily detected in the form of oedema and blebs between the myelin lamellae as vacuoles.

Synaptopathy

The presynaptic nerve endings possess vesicles in which the neurotransmitters are stored. The transmitter substance is passed on to the post-synaptic membrane containing receptors (specific-binding proteins) by diffusion. The binding of the transmitter to these receptors cause excitation. Synaptopathy results in impaired neurotransmission at synaptic regions due to the interference of xenobiotics with the biosynthesis and metabolism of the transmitter, axonal transport and release of the transmitter, receptor signal transmission, interaction of transmitter with the receptor or uptake of the transmitter. Agents like kainic acid, glutamate, phenoxybenzamine, boron hydroxide, carbon disulphide, pesticides (DDT and dieldrin), etc., are found to cause synaptopathy.

Neurotoxic Effects on Electrical Properties of Neuron

Conduction and processing of information constitute electrical phenomena which are carried out by the neurons. The nervous system is concerned with each and every activity of the animals so that the neurotoxins affect a wide range of functions through alterations in the electrical properties of the neurons either indirectly or directly.

Indirect effects   These include changes in the conductive properties due to structural damages in the neurons (cell bodies) and myelin sheath and impaired cellular transport mechanism. Mechanical or chemical damage in the nerve fibres would cause swelling and constriction of the fibre resulting in disturbed conduction of nerve impulses like decreased conduction rate, complete blockage of action potential, and so on. The neural membrane is very thin and separates two conductive fluids and so its capacity is very limited. Therefore, any current which passes through the membrane has to charge or discharge this capacity. So, the membrane at the swelling point takes longer time to depolarize (delayed conduction). The delay caused by the swelling may be so great that enables the membrane to be excitable again with depolarizing current. Now the action potential spreads back in the direction from which it arose. This phenomenon is called **repetitive firing** or **reflection.** Acrylamide, isoniazid, 2,5-hexanedione, IDPN, vinca alkaloids, etc. are found to cause conductive disorders through swelling. The disturbed conduction may also be due to changes caused by the neurotoxins in the composition of ion channels found in the membrane. In this case, the sensory fibres are first affected followed by the motor fibres. For example, doxorubicin causes loss of function in the sensory ganglia but does not affect the motor cells. On the other hand, organophosphates are shown to damage the motor cells first followed by sensory neurons (delayed neuropathy).

The substances which affect the Schwann cells which are responsible for the formation of myelin sheath are called as **myelinotoxins.** They reduce the action of the myelin sheath with diminished conduction rate but without entire blockage of conduction. Due to swelling, the insulating effect of the myelin sheath is prevented and so on axon may stimulate the adjacent one (cross talk). Hexachlorophene, isoniazid, cyanate, triethyltin, pyrithiamine, carbon monoxide, cyanide, lead, etc. are some examples of myelinotoxins. The effect of neurotoxins on the cell bodies are more or less similar to those that cause axon degeneration.

The process of degeneration may start at the cell body and may proceed towards its ending (dying forward) and the opposite phenomenon is called "dying backward" in which the axon degenerates at their ends and progresses towards the cell body. Certain toxicants reach the central nervous system from the periphery through neuronal transport where the substance penetrates the nerve endings at the periphery as in the case of tetanus toxin.

Direct effects   The excitable ion channels in the nerve membrane are of two types namely, **Na+** and **K+ ion channels.** These channels consist of specialized membrane proteins which selectively allow the ions to pass through the membrane.

Na+ ion channel   This channel is responsible for quick depolarization during action potential. The Na+ channel protein is a very complex molecule, so a variety of neurotoxins cause a wide range of effects leading to depolarization. It is a large protein containing a pore which allows Na+ ions selectively and two closing mechanisms, namely, the m-gate and the h-gate as in Figure 18.2. The m-gate is closed during resting membrane potential and opens rapidly during depolarization (sodium activation). The h-gate opens during resting state and closes slowly during depolarization (sodium inactivation). Thus opening of the m-gate initiates the sodium flow and the h-gate initiates the repolarization. Tetrodotoxin, saxitoxin, trimethyl oxonium, etc. block the Na+ ions to pass through the membrane from outside, thus reducing the permeability of the Na+ ions. N-bromoacetamide iodate and H+ ions prevent the closure of the h-gate inside the membrane. Sea anemone toxins and scorpion venom exhibit the same effect from outside the membrane. Batrachotoxin, grayanotoxin, veratridine and insecticides like DDT and pyrethroids prevent the closure of the m-gate.

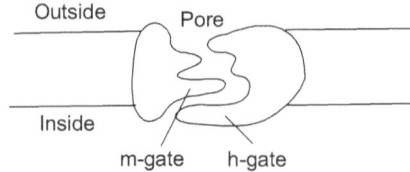

Figure 18.2   Structure of Na$^+$ channel protein

K$^+$ ion channel   This channel includes several types of which some are located on the nerve endings and nerve cells and others are found on the axons. In axonic channels, neurotoxins exert no selective action. The substances having anaesthetic effect (volatile organic solvents) are found to block the functioning of K$^+$ ion channel.

In addition to the above channels, Ca^{++} ion channel also plays an important role in linking the electrical membrane phenomena with other processes. For example, the Ca^{++} ion channel in the muscle cells links the electrical membrane phenomena with contraction and in the neuron with the secretary processes. Divalent ions such as nickel, lead, magnesium, cobalt, cadmium, etc. are capable of blocking the Ca^{++} ion channel.

NEPHROTOXICITY

Nephrosis refers to the degeneration of kidneys and it may be **glomerulonephrosis** or **tubulonephrosis.** The disorders of glomerulus are collectively called as **glomerulopathies** which may be due to either degeneration of glomerulus without inflammatory changes (**glomerulonephrosis**) or due to inflammatory reactions (**glomerulonephritides**). The glomerulopathy also includes inflammation and fibrosis in the renal interstitial cells (interstitial nephritis). These alterations will result in urinary stasis in the nephron, distension of Bowman's capsule and atrophy as well as sclerosis of the glomeruli. The kidney is enlarged and may contain "amyloid", a hyaline material in **renal amyloidosis**. The proximal tubule is often affected by nephrotoxins which cause degeneration of tubules followed by inflammatory changes. During the initial stages of this process, hydropic degeneration occurs resulting in necrosis. Renal papillary necrosis and deposition of crystals are encountered by Henle's loop, the distal convoluted tubule, and collecting tubule. The renal damage would affect the two important functions of the kidneys, namely, the formation of urine and the reabsorption of substances. Therefore, the composition of urine is altered so that the urine may contain more water (**polyuria**) or less water **(oliguria)**. The reduced absorptive capacity is indicated by the excretion of protein in the urine **(proteinuria)**. Table 18.11 gives an account of various nephrotoxic disorders caused by different toxicants.

Agents that Cause Renal Tumours

Opportunistic bacteria are found to produce nitrosamines in man followed by the infection of *Schistosoma haematobium* and to cause cancer in the urinary bladder. A number of substances like N-naphthalamine, methylnitrosourease, etc. are shown to be the initiators of carcinogenesis whereas chemicals like D, L-tryptophan, saccharine, etc. are the promoters of carcinogenesis in the urinary bladder of experimental animals such as dogs and rats. Immunosuppressants like cyclophosphamide are shown to be carcinogenic resulting in the urinary bladder cancer in man. An analgesic phenacetin is a stimulant of chronic degenerative kidney disorder causing cancer in the urinary tract and renal pelvis. Polycyclic aromatic amines such as nitrofurans and nitroimidazoles induce bladder cancer in rats.

Table 18.11   Nephrotoxic effects caused by nephrotoxicants

Nephrotoxic substances	Nephrotoxic disorders
Mercury (mercuric chloride and dimethyl mercury)	Glomerulonephropathies, proteinuria and deposition of immune complexes.
Platinum (cisplatin)	Damages in proximal and distal tubular cells, cortical and medullary lesions and atrophy of cortex.
Cadmium (cadmium chloride)	Damage in tubules and proteinuria.
Haloalkanes (carbon tetrachloride, chloroform) and haloalkenes (hexachlorobutadiene, dichlorodifluoroethane)	Renal damage and impairment of renal function.
Antibiotics	
Aminoglycosides	Nephrotoxic side effects such as changes in proximal tubular cells and endothelium of glomerulus.
Cephalosporins	Reduced tubular excretion.
Analgesics (cyclosporin A)	Renal damage, loss of renal papilla, inflammation of interstitial cells, fibrosis, symptoms of renal injury and nephrotoxic side effects.

REPRODUCTIVE TOXICITY

Reproductive toxicology deals with the study of mechanisms and effects caused by the toxicants at different stages in the reproductive cycle of male and female animals. The harmful effects due to the toxic substances in the male and female reproductive functions or in the progeny are referred to as **reproductive toxicity**. While teratology deals with structural birth defects, the developmental toxicity is concerned with the effects caused by the toxicants prior to attaining adult life. The reproductive toxicity may manifest throughout the entire reproductive cycle based on the sensitivity of the stage to various agents. This sensitivity mainly depends on the genotype of the fertilized egg, its development and dose of the substance. The toxicants interfere with the reproduction and development through mutation, chromosomal aberrations, disturbances in cell division, changes in DNA and protein synthesis, reduction in the raw materials for energy and enzyme production and altered membrane functioning with the resultant osmotic and ionic balance. These changes would cause cell death, disturbed cell contact and biosynthesis and disruption of tissue structure during embryogenesis.

The reproductive toxicity may lead to functional and fertility disorders, malformations, retardation of growth and intrauterine death. The functional disorders are seen during the foetal period and later stages of development and include metabolic, immunological and reproductive alterations which manifest during the formation and differentiation of organs and organ systems. The fertility disorders include reduced gametogenesis and sexual potency, interference with sex hormones, production of abnormal sperms in males causing sterility and altered ovulation, menstrual cycle, fertilization, implantation, pregnancy period, parturition, lactation, and so on in females. The structural defects occur mainly in embryonic stages and are used to evaluate teratogenicity of the toxicants. A recognizable pattern of malformations is referred to as **syndrome**. For example, the foetal alcohol syndrome indicates the defects in the craniofacial region, central nervous system and cardiovascular system. A malformation along with the resultant structural changes is called **anomaly**. Encephaly is an anomaly in which there is a complete absence of brain tissue along with abnormalities in ears, eyes and neck. A recognizable malformation other than syndrome and anomaly is called **association,** an example of which is the VATER association which is characterized by abnormalities in vertebrae, anus, trachea, oesophagus and kidneys. Growth retardation may occur in the foetus with alterations in the body or organ weight of the foetus. The intrauterine death is mainly due to infections, chemicals, diet and chromosomal aberrations and may occur at the period between fertilization and implantation or at the time of implantation or after implantation.

The intake of drugs such as narcotics, sleeping pills, anti-hypertensive sulphonamides, etc. by the mother would cause pharmacological effects in the neonates. The embryonic and foetal tissues are highly sensitive to carcinogens. Substances like nitroso compounds, polycyclic aromatic hydrocarbons and mycotoxins are found to cause transplacental carcinogens in the offsprings of the female experimental animals exposed to these compounds during pregnancy. The toxicants thus cause adverse effects on the functioning of the reproductive system at various stages of the reproductive cycle and an overall picture of these substances and their effects with reference to humans are explained in Table 18.12.

Table 18.12 Effects of toxicants in male and female reproductive systems of man during various stages of reproductive cycle

Developmental stage	Affected organs/functions		Effects
	Man	**Woman**	
Gametogenesis	Spermatogenesis	Oogenesis	Sterility, damaged sperms and ova, chromosomal aberrations and hormonal imbalance.
Fertilization	Secondary reproductive organs	Oviduct	Impotency, sterility and chromosomal aberrations.
Implantation	–	Uterus	Abortion, resorption of foetus, chromosomal aberrations and low birth weight.
Embryogenesis	–	Uterus	Abortion, foetal death, congenital abnormalities, chromosomal aberrations and low birth weight.

(Contd.)

Table 18.12   [Continued]

Developmental stage	Affected organs/functions		Effects
	Man	Woman	
Organogenesis	-	Placenta	Congenital abnormalities, abortion, foetal death, chromosomal aberrations, retardation of growth and development, and transplacentational carcinoma.
Foetal development	-	Foetus	Premature birth, congenital abnormalities, death at birth and low birth weight.
Postnatal	-	Child	Infant mortality, mental retardation and metabolic, functional and developmental disorders.

ENDOCRINE TOXICITY

The endocrine system is highly sensitive to the toxicants as the hormones secreted by the endocrine glands maintain an internal homeostasis in the body through the regulation of various metabolic processes.

Table 18.13   Toxicants that affect the endocrine glands and their effects

Toxic substance	Target organ	Effects
Thiocyanate, thiomides, lithium and sodium bromide, TBTO, etc.	Thyroid gland	Inhibition of synthesis of thyroid hormones (hypothyroidism), and hyperthyroidism.
Antimicrobial veterinary drug (carbadox)	Adrenal cortex	Decreased aldosterone.
Nicotine, thiouracil, reserpine, etc.	Adrenal medulla	Hyperplasia and pheochromocytomas.
Cannabimids, lindane, *n*-hexane, phthalate esters, sodium bromide, etc.	Testis	Reduced number and reduced steroids in Leydig cells, testicular atrophy and reduced spermatogenesis.
Insecticides	Ovary	Arrested ovulation, reduced reproductive cycle and disturbed reproductive function.

Toxicants may have a direct hormonal effect by interacting with the hormone receptors or an indirect effect through altered biosynthesis, uptake, release, metabolic conversion and excretion of hormones. The toxic impact of a number of substance on endocrine system have been evaluated through *in vitro* and *in vivo* studies. The endocrine glands such as thyroid gland, pancreas, adrenal gland and gonads are more frequently used to assess the effects of toxic substances. Through experimental studies, the xenobiotics are shown to cause impaired growth or decreased metabolism due to reduced thyroid function, changes in sugar metabolism due to disturbed pancreatic function, altered renal function due to adrenal dysfunction and reduced fertility due to abnormal gonadial function. It is also shown that the toxic substances induce endocrine organs such as the pituitary gland, thyroid gland, pancreas and gonads. The increased hormonal stimulation may produce cellular deviations leading to tumours or may induce the growth of tumour cells in target organs.

In Table 18.13 various substances that bring toxic effect on the endocrine glands of experimental animals are shown.

IMMUNOTOXICITY

Immunotoxicity refers to the events of unwanted effects due to the interaction between the toxic substances and the immune system. The immune system operates by producing different kinds of cells which are of great importance for the efficacy of the system. These immunocompetent cells proliferate and differentiate in order to evoke immune response. Thus, the immune system is highly organized and complex as well as extremely well balanced with continuous intensive communication between variable cell types. In an immunological process, the antigen which enters into the body is recognized with the resultant proliferation and differentiation of lymphocytes followed by clonal expansion. Now the B cells produce antibodies and T cells produce cytokines to react with the foreign substances. The immunotoxic substances interfere with this homeostatic system resulting in altered immune functions, namely, **immunosuppression** (deficient resistance) and **immunostimulation** (autoimmunity, hypersensitivity).

Target Sites of Immunotoxins

The exact mechanism of immunotoxic effects and physiological functioning at the immune system are not fully known. Figure 18.3 shows the important components of the immune system that are liable to the effects of various toxicants.

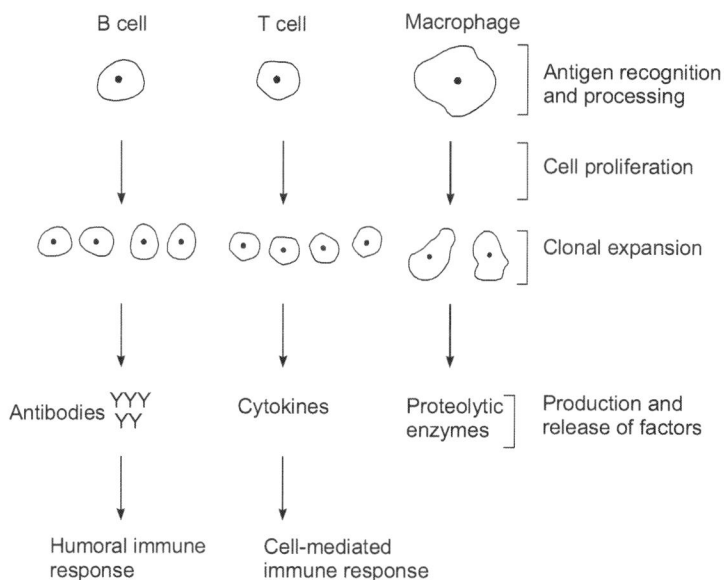

Figure 18.3 Immunological system that are sensitive to the toxicants

In short the immunological process involves four phases, namely, the phase of antigen recognition in which the lymphocytes come in contact with the antigen, the adaptive phase in which the lymphocytes undergo proliferation, differentiation phase (clonal expansion) in which a large number of lymphocytes are produced and manifestation phase in which antibodies, cytokines and proteolytic enzymes are produced by the B cells, T cells and macrophages respectively. As proliferative activity of the cells is high in the above phases, the immune response is very sensitive to cytotoxic substances. Therefore, certain toxins exert anti-proliferative effects in the immune system. A number of substances adversely affect the immune system indirectly via the endocrine system.

Immunotoxic Substances

In recent times, there is an increased incidence of infections and other disorders in human due to immunotoxicants. Several chemicals that occur in the environment such as polychlorinated biphenyls, polybrominated biphenyls, hexochlorobenzene, lead, cadmium, methyl mercury, benzopyrene, benzidene, nitrogen dioxide, asbestos, dimethyl-nitrosamine, air-polluting gases and aerosols, etc. are found to affect the immune system in experimental animals as well as in human beings. As far as pharmaceuticals are concerned, the immunosuppressants such as glucocorticoids or azathioprine which are used to prevent rejection in transplantations, cause increased infections and neoplastic disorders like lymphomas and leukaemias. In Table 18.14, major groups of immunotoxic substances and their impact on immune system in experimental animals are shown.

Table 18.14 Immunotoxins and their effects in experimental animals

Toxic substance	Effects
Organotin compounds	Reduction in number of lymphocytes and weight of spleen and thymus, thymic atrophy, inhibition of growth of immune cells, reduction of lymphocytes in lymph nodes and spleen, increased IgM and decreased IgG antibodies, delayed-type hypersensitivity, reduced proliferation in thymus and spleen and reduced activity of NK cells.
Polychlorinated biphenyls	Thymic atrophy, suppression of humoral response, reduced T cell activity, disturbed phagocytosis by macrophages and cytotoxicity by NK cells and reduced resistant to infections.
Dibenzofurans	Thymic atrophy, decreased IgA and IgM levels, reduced number of T_H cells and reduced delayed-type hypersensitivity.
Dibenzo p-dioxins	Thymic atrophy, reduced delayed-type hypersensitivity and decreased T_C cell activity.
Oxidizing gases	Resistance to respiratory infections, reduced NK cell activity, reduced phagocytic activity of macrophages, suppression of delayed-type hypersensitivity and increased sensitivity to respiratory infections.

Substances that Cause Immunostimulation

While the above mentioned immunotoxins suppress the activity of the immune system, certain substances are found to evoke systemic allergic reactions and autoimmune disorders due to inappropriate functioning of the system. In conditions of allergy, the elimination of allergens by the immune system produces undesirable side effects mainly in the form of inflammations. The most important allergic reactions are the immediate type hypersensitivity and delayed-type hypersensitivity. In these disorders, a toxic substance acts as a hapten and reacts with the endogenous compounds (proteins) so that the endogenous compounds become foreign substances to the immune system. Therefore antibodies are produced against these components. Antidepressants (zimeldine), anaesthetics (halothane), vinyl chloride, chloromazine, isomazite, some anticonvulsants, etc. are found to cause allergic and autoimmune reactions. Hexachlorobenzene is a potential stimulation of the immune system. It causes increase in weight of lymphoid organs, hyperplasia in lymphoid tissue, proliferation of the high endothelial venules, accumulation of macrophages, increased IgM level, increased neutrophils, basophils and mononuclear cells, and so on. It also simulates the synthesis of antibodies against thymus-dependent antigens and also delayed-type hypersensitivity reactions.

Determination of Immunotoxicity

As immunotoxic substances are being continuously increasing, it is not possible to test all the substances for their potential effects on all the aspects of the immune system. In general, screening of the immunotoxicity has to be incorporated into routine toxicity studies and then the effects on immune system and defence mechanism are to be established. In immunotoxicological studies, it is necessary to evaluate whether the changes in immune function are due to direct effects or due to indirect effects resulting from disturbed protein synthesis, stress, hormonal imbalance, malnutrition, etc. As the immune system is highly susceptible to a number of toxic substances, the immunological assays have to be carried out through hierarchial system of panels, namely, primary immunotoxicity screening panel and secondary immunotoxicity screening panel. The primary panel involves immunopathology as a part of general toxicology and includes evaluation of leucocytes differentiation, serum immunoglobulin levels, cellularity of bone marrow, weight and histology of lymphoid organs, and so on. The second panel includes the immune function test based on the results obtained from the primary panel.

Identification of Defence Mechanisms

The non-specific defence mechanisms play a vital role as "front-line defence" prior to the onset of a specific immune response. Phagocytosis and lysis are the two important aspects of non-specific defence mechanisms and can be tested by both *in vitro* and *in vivo* studies. Cytotoxicity by the macrophages or NK cells is another non-specific defence mechanism and can be determined by using radioactive elements. The extent of proliferation of sensitized lymphocyes by specific antigen like ovalbumin correlates delayed-type hypersensitivity and can be measured by the incorporation of radioactive thymidine into DNA. The ability of proliferation of T cells can be tested *in vitro* through the incorporation of T-cell mitogens. In general, the rejection of allografts which is T-cell mediated, can be used to determine the immunotoxicity. The humoral immunity may be T-cell-dependent or T-cell-independent. The T-dependent antibody production can be determined by using T-cell-dependent antigens such as tetanus toxoid, sheep red blood cells, ovalbumin, etc. The antibody titres in the serum can be evaluated by enzyme-linked immunosorbent assay (ELISA) or by haemagglutination test. The synthesis of immunoglobulins by the plasma cells can be measured in immunized rats through the determination of antibody titers using ELISA technique. In recent times, *in vitro* studies have been given much attention to assess the substances for genotoxicity. The potential immunosuppressive ability of food additives, organometallic compounds and polychlorinated biphenyls can be determined by antibody production in a "Plague Forming Cell Assay" (*in vitro* immunization/Ab plague forming cell test.

REVIEW QUESTIONS

1. Write an essay on toxicants and their effects on organs/organ systems.

2. Explain irritant dermatitis and allergic dermatitis and their characteristics.

3. What is a photo-allergic reaction?

4. Write a brief account on different types of skin tumour and their causatives.

5. What are the major injuries in the gastrointestinal tract caused by the toxicants?

6. Categorize the gastrointestinal toxicants.

7. Describe acute and chronic disorders caused by hepatotoxins.

8. Explain the mechanisms of deposition of gases, vapours, and aerosols.

9. Give an account of direct and indirect effects caused by the inhaled particles and gases in the respiratory tract.

10. Describe the exposure or inhalation chambers and their use in respiratory function tests.

11. How could test atmosphere and particles be generated in inhalation chambers?

12. Explain various methods that are used to determine the particle size distribution of aerosols.

13. Enumerate the cardiotoxicants and their effects on the functioning of heart.

14. Write an account on malignant blood disorders and their causative agents.

15. Classify the neurotoxic effects based on the location in the nervous system.

16. Describe the structural and functional alterations caused by the neurotoxins.

17. Define the following terms and explain their characteristics and causative agents:

    i. Axonopathy

    ii. Myelinopathy

    iii. Synaptopathy

18. How could neurotoxins affect the electrical properties of neurons?

19. Describe the mode of action of neurotoxins on the excitable ion channel present in the nerve membrane.

20. Write a note on Na^+ ion channel K^+ ion channel and Ca^{++} ion channel.

21. What are the important categories of nephrotoxins?

22. In what ways nephrotoxins affect the excretory organs?

23. Write an essay on nephrotoxins and their effects.

24. Give an account of the agents that cause renal tumours.

25. Explain the various functional and fertility disorders caused by toxic substances.

26. How could the toxicants affect the reproductive functions during various stages of reproductive cycle in man?

27. Enumerate the toxicants that affect the endocrine glands and their effects in experimental animals.

28. Write an essay on immunotoxic substances.

29. What are the target sites of immunotoxins?

30. Describe the substances that induce immunostimulation.

31. Give an account of various methods used to determine immunotoxicity.

32. How could toxicants affect the non-specific defence mechanisms?

19

IMPACT OF POLLUTANTS ON AQUATIC ORGANISMS

Today, environmental pollution which causes hazards to fauna and flora as well as human health, has become a major global issue. Various pollutants cause greatly variable effects on animals. A wealth of information is available regarding the harmful effects of pesticides, heavy metals and industrial effluents on various animal groups, both invertebrates and vertebrates. The pollutants in water are shown to affect the feeding, food utilization, oxygen consumption, metabolic turnover, muscular action, endocrine coordination and enzyme action, as well as reproduction of aquatic organisms. Even at sublethal concentrations, the pollutants affect the life of aquatic fauna which are manifested as changes in physiology, biochemistry and activity levels of many enzymes.

PHYSIOLOGICAL EFFECTS

Changes in Digestive Physiology

In animals, the required energy is mainly derived from the foodstuff. Food selection in animals is often more influenced by taste receptors than by nutritive value. It has been evidenced that the pollutants affect the chemoreceptors both in invertebrates and fishes in such a way that they interfere with their feeding behaviour. The industrial effluents are found to diminish the food consumption and intestinal absorption in dragonfly larvae. Many histopathological alterations have been observed in the trichoid sensilla on the mouthparts of the larvae under the effluent toxicity. This would have disharmonized the electrophysiological responses on the activity of the receptor neuron of the sensilla resulting in the failure of sensory detection of food. The fall in the absorption of food could either be due to the inhibition of the activity levels of intestinal enzymes or due to the disruption of the brush border of the intestinal mucosa. These changes in the quantum of feeding and absorption would affect the normal physiological functions such as assimilation and metabolic rates in animals.

Changes in Food Utilization

Different food utilization parameters such as consumption, digestion, assimilation, conversion and excretion in animals are directly or indirectly affected by pollutants. The changes caused by the industrial effluents in the physiological functions like assimilation and metabolic rates could result in retarded growth. This implies that most of the energy could have been diverted to meet the energy demands to combat the toxic effects. Correlated with the decreased rate of growth and metabolism, the conversion efficiency could also reduce. The inhibition of body growth due to decreased food utilization would result in overall physiological disturbances in various tissues of the body.

Changes in Respiratory Physiology

Pollutants such as pesticides, heavy metals and industrial effluents are found to interfere with the respiratory metabolism of crustaceans, insects, bivalves and fishes causing a diminution in the rate of oxygen uptake. This could be due to histopathological changes in the respiratory surface area of animals under the toxicity of pollutants.

Changes in Haematological Parameters

Blood often exhibits pathological changes before the appearance of any external symptoms of toxicity. Therefore, the haematological studies in animals form a promising tool for the investigation of physiological alterations caused by the environmental pollutants. Many toxicants are found to inflict a reduction in the number of erythrocytes, and reticulocytes (Figure 19.1), haemoglobin content, leucocytes and PCV (packet cell volume) and MCHC (mean corpuscular haemoglobin concentration) and increased values of MCV (mean corpuscular volume) and MCHB (mean corpuscular haemoglobin) in the peripheral blood of fishes.

The decline in RBC number and Hb content could be due to either destruction of erythrocytes or inhibition of erythropoiesis or both. The declined MCHC values in the blood indicate the destruction of erythrocytes. Moreover, the pollutants cause significant reduction of different types of cells in the bone marrow of fishes, thus evidencing disturbed action of erythropoiesis in the bone marrow. This is supported by the fact that the platelets decrease in number during bone marrow depression. Many pesticides are found to cause reduction in number of RBCs and haemoglobin contents due to iron deficiency with resultant inhibition of haemoglobin synthesis. In erythrocytes of fishes, heavy metals are found to inhibit the biosynthesis of haemoglobin.

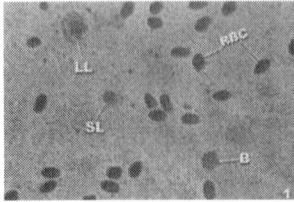

Blood smear of untreated fish

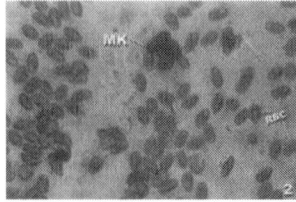

Blood smear of fish treated
in 0.02% effluent

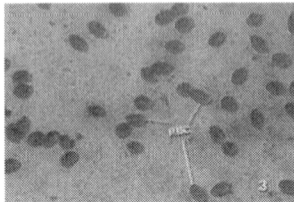

Blood smear of fish treated
in 0.04% effluent

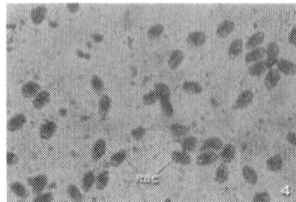

Blood smear of fish
in 0.06% effluent

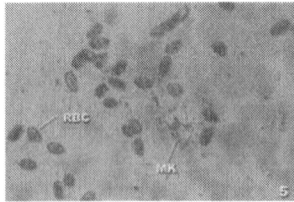

Blood smear of fish treated
in 0.08% effluent

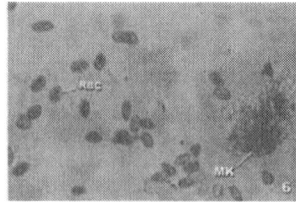

Blood smear of fish
in 0.10% effluent

RBC—Red blood corpuscle; LL—Large lymphocyte; SL—Small lymphocyte;
B—Basophil; MK—Megakaryocyte]

Figure 19.1   Blood smear in untreated and tannery effluent-treated
O. mossambicus

The disturbed bone marrow activity with abnormal erythropoiesis and fall in WBCs and reticulocytes as well as reduced Hb content would lead to a condition of macrocytic megaloblastic anaemia.

The WBCs are capable of phagocytosis and produce immune bodies which destroy foreign bodies as well as combat the toxicants introduced into the bloodstream. Insecticides and heavy metals are found to reduce the number of differential WBCs in freshwater fishes. The decreased number of neutrophils, lymphocytes and monocytes and an enhanced eosinophil number is count indicative of severe or initial stage of infection. In general, the pollutants could make the animals to lose their defensive mechanisms against infections.

Changes in Haemolymph Physiology of Insects

In insects, the haemolymph volume reflects the efficiency of transport of pressure from one part of the body to another and is determined by the water content. The concentration of water must be maintained within certain limits in the body for many physiological activities. The paper and pulp mill effluent is found to cause osmotic imbalance in the odonate larvae resulting in the decrease of haemolymph volume. This in turn could lead to a marked reduction of body weight.

The regulation of water and salt balance in insects is mainly concerned with maintaining a constant composition of the haemolymph involving osmotic pressure and the concentration of inorganic and organic solutes. Industrial effluents are shown to cause a decrease in the levels of Na^+, K^+ and Ca^{2+} in the haemolymph of insects. This could be due to the movement of solutes causing large changes in the haemolymph volume and osmotic pressure.

Many poisonous chemicals, plant extracts and industrial effluents are reported to bring about a significant reduction of total haemocytes in general and plasmatocytes and granulocytes in particular in the haemolymph of dragonfly larvae. The plasmatocytes are found to be involved in immune response and the phagocytes in phagocytosis with well-developed vacuolar apparatus. Therefore, the reduction in number of plasmatocytes and granulocytes would block the phagocytic activity, so the insect will become more susceptible to the toxicity as well as to infections. In insects, haemolymph coagulation is the initial step in wound healing involving coagulocytes and occasionally granulocytes.

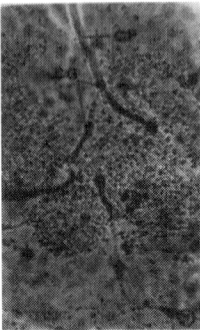

Haemolymph coagulation in the larva treated in 2% effluent (400×)

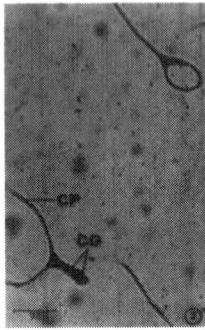

Haemolymph coagulation in the larva treated in 4% effluent (400×)

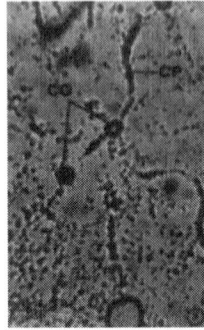

Haemolymph coagulation in the larva treated in 6% effluent (400×)

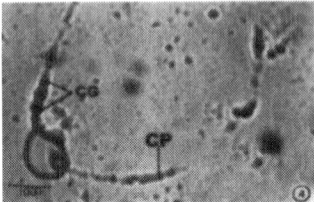

Haemolymph coagulation in the larva treated in 8% effluent (400×)

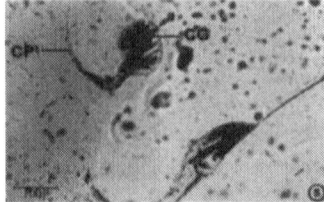

Haemolymph coagulation in the larva treated in 10% effluent (400×)

[CG—Coagulocyte; CP—Cytoplasmic process]

Figure 19.2   Impact of paper-pulp mill effluents on the coagulation of haemolymph in the larva of dragonfly *B. geminata*

The haemocytes are found to synthesize and secrete glyco- and lipoproteins which form matrices of haemolymph coagulation. These events are shown to be inhibited in the dragonfly larvae treated with paper and pulp mill effluent (Figure 19.2), which would thus impair the coagulation process of the haemolymph. Therefore, the insect would become susceptible to haemolymph loss in the event of wounding.

Changes in Reproductive Physiology

In dragonflies, the tannery effluent is found to cause vacuolization in the egg cytoplasm and nucleoplasm. The follicular epithelial cells are compactly arranged with very narrow spaces. They show infrequent cell divisions with the resultant cells having irregular shape. There is no sign for the initiation of vitellogenesis thus showing the adverse effects of the effluent on the normal development and maturation of oocytes in odonates. Moreover, many toxic substances are reported to affect fecundity in many insect species. Heavy metals are shown not only to affect the growth and development but also hatchability of many insect larvae. The insecticides and industrial effluents reduce the size of testis and inhibit spermatogenesis with the resultant reduction in number of deformed sperm bundles in insects.

BIOCHEMICAL EFFECTS

The animals try to overcome toxic stress by exhausting the organic reserves which become reduced in vital tissues leading to a severe physio-metabolic dysfunction. Decreased free amino acids and proteins in different tissues of animals could be attributed to the utilization of the same in various catabolic reactions under emergency. Under the conditions of decreased oxygen uptake, the animal has to depend upon anaerobic utilization of sugar which would culminate in the accumulation of lactic acid. The decrement of non-reducing sugars and glycogen content in the tissues indicate that these reserves are subjected to glycogenolysis to meet the energy demands during stress situation attributing to the prevalence of anaerobic condition. The depletion of carbohydrates as a whole along with proteins and amino acids indicates the operation of glycogenesis in order to mitigate the toxic stress. As a result, the animal is forced to depend on the energy from the metabolism of triglycerides from adipose tissue and amino acids from tissue proteins. The triglycerides are broken down into glycerol and free fatty acids which might actively participate in energy production through glycolytic and β-oxidation pathways respectively. The decreased cholesterol content is suggestive of lesser association of phospholipids in tissues.

Changes in the Activity Levels of Enzymes

Whenever there is a shift in the quantity or nature of fuel supply, the enzymatic constitution changes so as to make most efficient use of materials at hand. The exposure of animals to toxicants for even a short period may produce considerable destruction of their enzymatic architecture.

Pollutant stress is shown to increase the activities of various enzymes such as protease, lactic dehydrogenase (LDH), acid and alkaline phosphatases and lipases in the body tissues of animals. At the same time, the tissues exhibit a decreased rate of succinic dehydrogenease (SDH) activity. The suppression of SDH activity indicates the interference of the toxicant in the oxidative metabolic pathway of the animals and it is supported by elevated LDH activity. The decreased activity of SDH under toxic effect could be considered as an indication of shifting towards anaerobic metabolism. Lactic acid is the end product of anaerobic metabolism and is mobilized for oxidation to meet the energy demands, as the LDH activity accelerates the conversion of lactic acid into pyruvic acid. The increased LDH activity enables the tissues to maintain the TCA cycle in a functional state so that the tissues derive maximum energy from carbohydrate metabolism. Rise in the tissue lipase activity would stimulate intracellular lipolysis in accordance with energy demands of the animal during stress. The increased acid phosphatase activity could result in the rupture of cellular and lysosomal membranes showing that the toxicants have a strong impact on the degradation of cells in the body. The enhancement of the activities of phosphatases may also be due to the tissue inflammatory reactions of toxins or due to increased transphosphorylation activities of the tissues.

Many toxicants like pesticides and heavy metals inhibit the activity of acetyl choline esterase (AChE) with resultant accumulation of acetylcholine (ACh) in the tissues of animals. Pesticides are also known to bind with the active sites of AChE preventing the breakdown of ACh. As a result, the conduction of nerve impulses in the synaptic regions of the nerves is disturbed. Thus, the death of the animal by the toxicity of a toxicant could be due to the inhibition of AChE in the nervous tissue and the consequent disruption of nervous activity by the accumulation of ACh. In other words, ACh may no longer be decomposed in the event of decreased levels of AChE so that the nerve would start firing in an uncontrolled manner resulting in the mortality of the animal.

In the tissues of pesticide-treated fishes, both aspartate and alanine amino transferases (AAT and ALAT) are found to be elevated confirming the stress condition. However, the per cent increase of AAT activity is found to be greater than ALAT indicating increased energy demands during toxic stress. In fishes, pesticides cause a significant decrease in glutamate dehydrogenase (GDH) in tissues. This is suggestive of accumulation of glutamine which is fed into the TCA cycle.

HISTOLOGICAL CHANGES

Sense Organs

In insects, the receptors found in the mouth parts are typical trichoid sensilla, which are contact receptors. The dendrites along with the scolopale and perikaryon of these sensilla exhibit structural deformities of varying magnitudes such as separation of dendrite from the scolopale resulting in wide gaps, disintegration of dendrites, and so on, under the toxicity of paper and pulp mill effluent (Figures 19.3 and 19.4).

Gut Wall

In animals, the midgut and hindgut are more prone to action of toxicants when compared to foregut. In the midgut, the apical regions of epithelial cells exhibit necrosis and vacuolization. The gut's epithelium has undergone fusion with resultant spaces amidst the cells. The epithelium separates from the muscle strands followed by degeneration of nidi cells, peritrophic membrane and muscular layers (Figure 19.5). The hindgut also exhibits similar histopathological alterations such as fusion of epithelial cells, separation of epithelium from the muscular layers and degeneration of cytoplasm of the cells.

Liver

The toxicants cause swelling of hepatocytes, dilation and congestion of sinusoids, coagulation of blood in sinusoids, cytoplasmic degeneration and nuclear disarray, extrusion of nuclei from hepatocytes and clotting of blood in sinuses, etc. in the liver of animals.

Gill

The histopathological changes observed in the gills of fishes include mucous coating on the surface and blackening of gill filaments, rupture of capillaries, bulging of tips of gills, degeneration of gill filaments, necrosis of gill epithelium and separation of gill filaments from the axis.

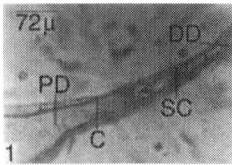

Longitudinal section of sensilla in untreated larva 450×

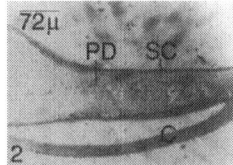

Longitudinal section of sensilla in the larva treated in 0.5% effluent 450×

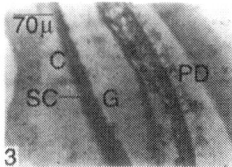

Longitudinal section of sensilla in the larva treated in 1.0% effluent (675×)

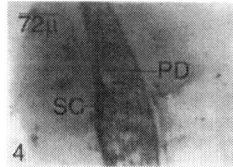

Longitudinal section of sensilla in the larva treated in 1.5% (450×)

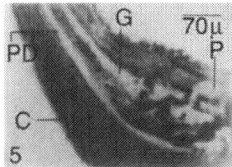

Longitudinal section of sensilla in the larva treated in 2.0% effluent (675×)

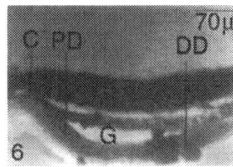

Longitudinal section of sensilla in the larva treated in 2.5% effluent (675×)

[C—Cuticle; DD—Distal dendrite; G—Gap; P—Perikaryon; PD—Proximal dendrite; SC—Scolopale]

Figure 19.3 Impact of paper and pulp mill effluent on trichoid sensilla of maxilla in the larva of dragonfly *M. cingulata*

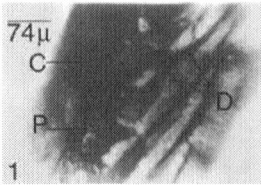

Longitudinal section of sensilla in untreated larva (150×)

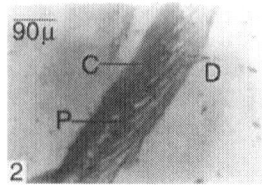

Longitudinal section of sensilla in the larva treated in 0.5% effluent (100×)

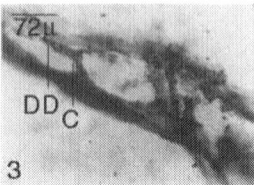

Longitudinal section of sensilla in the larva treated in 1.0% effluent (450×)

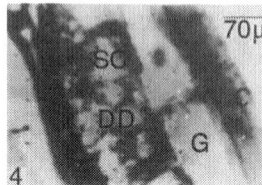

Longitudinal section of sensilla in the larva treated in 1.5% effluent (675×)

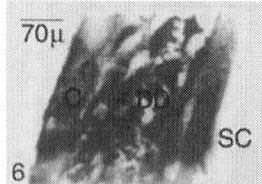

Longitudinal section of sensilla in the larva treated in 1.5% effluent (675×)

Longitudinal section of sensilla in the larva treated in 2.0% effluent (675×)

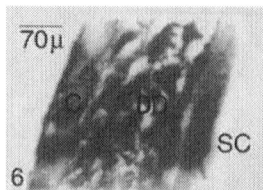

Longitudinal section of sensilla in the larva treated in 2.5% effluent (675×)

[C—Cuticle; D—Dendrite; DD—Distal dendrite; G—Gap; P—Perikaryon; SC—Scolopale]

Figure 19.4   Impact of paper-pulp mill effluent on trichoid sensilla of labium in the larva of dragonfly *M. cingulata*

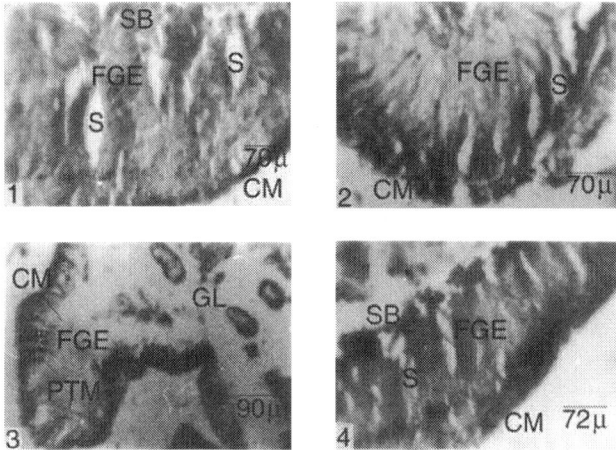

Transverse section of portions of midgut of the larva treated in 2.5% effluent at different magnifications

[CM—Circular muscle; FGE—Fused gill epithelium; GL—Gut lumen; PTM—Peritrophic membrane; S—Space; SB—Striated border]

Figure 19.5   Impact of paper-pulp mill effluent on the midgut in the larva of dragonfly *M. cingulata*

Rectal Gill

In dragonfly larvae, the industrial effluents are shown to cause oedema and necrosis in the gill leaflets with extensive vacuolization. The haemocytes accumulate at the gill bases, which become more swollen and congested with haemocytes. Ulceration and hyperplasia of gill leaflets are seen in addition to the fusion of adjacent leaflets (Figure 19.6).

Haemopoietic System

In fishes, different types of cells in the bone marrow are recorded to exhibit significant reduction, thus evidencing disturbed action of erythropoiesis in the damaged bone marrow.

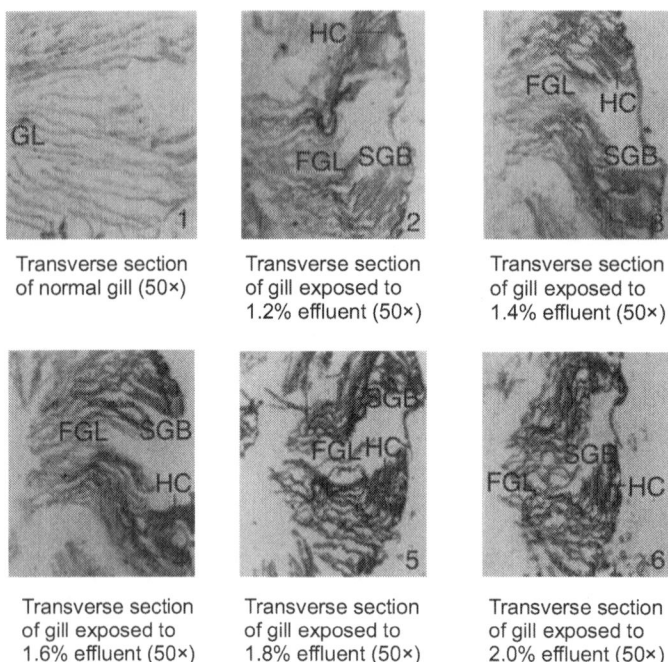

Transverse section of normal gill (50×)

Transverse section of gill exposed to 1.2% effluent (50×)

Transverse section of gill exposed to 1.4% effluent (50×)

Transverse section of gill exposed to 1.6% effluent (50×)

Transverse section of gill exposed to 1.8% effluent (50×)

Transverse section of gill exposed to 2.0% effluent (50×)

[GL—Gill lamellae; SGB—Swollen gill base; HC—Haemocytes; FGL—Fused gill lamellae]

Figure 19.6 Impact of tannery effluent on the rectal gills of the larva of dragonfly *P. flavescence*

In dragonfly larvae, the haemopoietic organ is well organized and situated in the 2nd and 3rd abdominal segments. Under paper mill toxicity, the cells of haemopoietic tissue have fused with resultant vacuolization. The cortex and medulla are not demarcated, and become vacuolated without differentiated haemocytes (Figure 19.7).

Kidney

Under toxic stress, the kidney of fishes exhibit vacuolization and necrosis of epithelial cells lining the renal tubules, dilation of glomerular capillaries, empty sinusoids with internal haemorrhage, degeneration of kidney tubules and swollen glomeruli.

Transverse section of a portion of tissue of the larva treated in 2% effluent (400×)

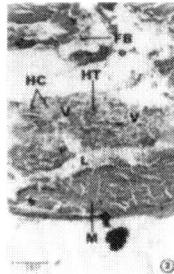

Transverse section of a portion of tissue of the larva treated in 4% effluent (400×)

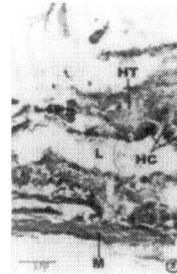

Transverse section of a portion of tissue of the larva treated in 6% effluent (400×)

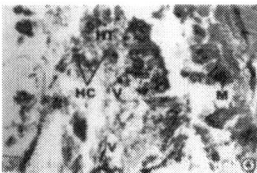

Transverse section of a portion of tissue of the larva treated in 8% effluent (400×)

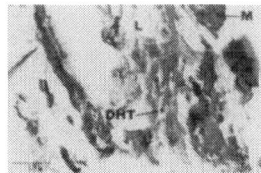

Transverse section of a portion of tissue of the larva treated in 10% effluent (400×)

[DHT—Deformed haemopoietic tissue; FB—Fat body; HC—Haemocytes; HT—Haemopoietic tissue; L—Lumen; M—Muscle; V—Vacuole]

Figure 19.7 Impact of paper and pulp mill effluent on the haemopoietic tissue of the larva of dragonfly *B. geminata*

Testis

Many insecticides are found to cause necrosis of sertoli cells, shrinkage and necrosis of seminiferous tubules and nuclear disarray in Leydig cells in the testis of many piscean species.

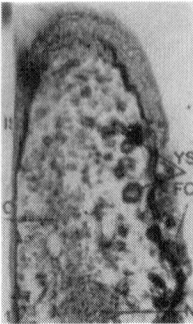

Oocyte with yolk bodies
at the periphery (400×)

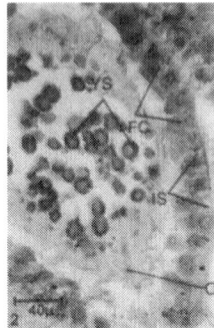

Oocyte with large dense yolk
bodies moving towards centre (400×)

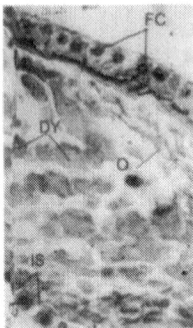

Oocyte with concentrated
base yolk bodies (400×)

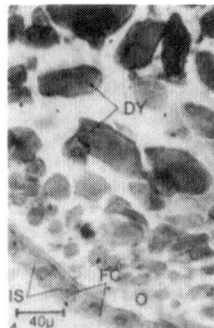

Oocyte with compact dense yolk
bodies covered with squamous
follicular epithelium (400×)

[DY—Dense yolk; FC—Follicular cell; IS—Intercellular space; N—Nucleus; O—Ooplasm; YS—Yolk spheres]

Figure 19.8   L.S. of normal vitellogenic basal oocyte in the dragonfly
P. flavescens

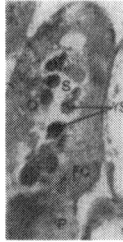

Oocyte treated in
0.5% effluent (400×)

Oocyte treated in
1.0% effluent (400×)

Oocyte treated in
1.5% effluent (400×)

Oocyte treated in
2.0% effluent (400×)

Oocyte treated in
2.5% effluent (400×)

[BO—Basal oocyte; DN—Disintegrated nucleus; FC—Follicular cell; NI—Nucleolus;
O—Ooplasm; P—Pedicel; S—Space; V—Vacuole; VN—Vacuolated nucleoplasm;
VO—Vacuolated ooplasm; YS—Yolk sphere]

Figure 19.9   L.S. of basal oocyte from the ovary in tannery effluent-treated
larva of dragonfly *P. flavescens*

Ovary

The industrial effluent is shown to bring about many histological alterations in oocytes of developing ovaries of odonates. Under the toxic stress, the basal oocytes become mostly oval in shape. The epithelial cells of the oocyte are more compactly arranged without intercellular spaces with vacuolated ooplasm and nucleoplasm. The nucleus is degenerated without nucleolus and nuclear membrane. The ooplasm contains clear spaces with very few yolk spheres indicating the absence of vitellogenesis (yolk formation) in the oocytes (Figures 19.8 and 19.9).

REVIEW QUESTIONS

1. Explain the harmful effects of pollutants on digestive physiology and food utilization in aquatic organisms.

2. How do the pollutants interfere with haematological parameters and respiratory physiology of aquatic animals?

3. How do the pollutants affect the haemolymph physiology of insects?

4. Give a brief account of alterations caused by the toxic substances on the reproductive physiology of insects.

5. Discuss how many pollutants at as metabolic stressors.

6. Describe in detail the changes in the activity levels of enzymes under toxicant stress.

7. Write an essay on the histopathological changes caused by the toxic chemicals on various tissues of animals.

20

ENVIRONMENTAL TOXICOLOGY

Environmental toxicology, which is a new and very broad discipline, is mainly concerned with the study of harmful effects of xenobiotics, present in the biosphere, on living organisms including man. This field is a multidisciplinary one and includes the school of chemistry, life sciences, agriculture, public health, etc. In other words, it is the science of chemicals that cause toxic effects on biota and alter the ecological balance of the environment resulting in pollution. The **environmental pollutants** are those substances which produce harmful effects on the environment as well as on living organisms. The **environmental contaminants** are those materials that are introduced into the environment by the activities of man leading to the deterioration of the environment. The environmental toxicants are the toxic substances found in the environment causing adverse effects on living organisms. In short, this field includes the application of toxicology to organisms in relation to the environment.

ENVIRONMENTAL POLLUTANTS

Any agent that deteriorates the natural environment is called environmental pollutant. The pollutants which naturally occur in an environment in greater quantity cause adverse effects on the environment, which in turn adversely affect the biota including humans. Thus an environmental pollutant can be defined as a substance that adversely affects the natural environment through the interference with the ecosystem or with the health and recreational values of man. The pollutants may either be non-degradable or biodegradable.

In general, environmental pollution includes water pollution, air pollution, land pollution, noise pollution, radioactive pollution, and so on, the sources of which may be natural or man-made. The major water pollutants are heavy metals such as lead, mercury, cadmium, chromium, arsenic, etc.; pesticides including synthetic and natural products toxic organic substances such as polycyclic aromatic hydrocarbons, plastics, detergents, etc.; toxic inorganic substances such as cyanide, fluoride, ammonia, etc.; nutrients like nitrogen and phosphorus and pathogens such as viruses, bacteria, protozoans, worms, etc. The air pollutants mainly include gaseous pollutants such as SO_2, NO_2, CO, O_3, H_2S, NH_3, Cl_2, PAN, hydrocarbons, etc. and inorganic, organic and bioparticulates. The land pollutants result from open dumps of refuses, accumulation of toxic chemicals in the soil, leaching of toxic chemicals into the soil, and so on. The radioactive pollutants include naturally occurring radioisotopes and cosmic rays and the man-made sources such as atomic explosions, nuclear emissions and wastes from nuclear power plants, and use of radioactive chemicals in the laboratory. In addition, the substances such as cosmetics, food additives, drugs, sanitary chemicals, etc. also form domestic poisons.

ENVIRONMENTAL CONTAMINANTS

Rapid industrialization, technological advancements, and successful green revolution in both developed and developing countries have resulted in the introduction of a variety of toxic materials into the environment. Generally, the environmental contaminants are called xenobiotics. At present a large quantum of chemicals are used by man and are found to be environmental contaminants. They comprise inorganic and organic industrial chemicals, agrochemicals (pesticides and fertilizers), heavy metals, detergents, dyes, drugs, plastics, food additives, cosmetics, etc. These contaminants along with the transformed products of the environment are mainly concerned with human health.

In an environment, the physico-chemical quality along with biological interactions determine the distribution of the organisms which respond to the stress caused by the toxic contaminants. Heavy metals and pesticides persist for a long time in the environment and cause undesirable effects on the biota. Among the contaminants, heavy metals are considered to be serious pollutants as they persist in the environment, exert toxic impact even at low concentrations and incorporate into the food chain through accumulation in the body of organisms. In addition, the organic matter of solid wastes interact with the metal ions and form a number of complexes. The industrial effluents contain organic and inorganic substances, which contaminate mainly the aquatic environment.

ENVIRONMENTAL TOXICANTS

The toxicants are introduced into the ecosystem intentionally or accidently so that they injure the quality of the environment thereby making it unfavourable for organisms. The entry of toxicants into the ecosystem occurs through agricultural run-off, urban run-off, disposal of sediments, etc. (non-point sources) as well as through the discharges from various industries, disposal sites, waste treatment plants, and so on (point sources). At present, much importance has been given to materials resulting from man's activity such as pesticides, industrial chemicals, heavy metals, industrial effluents, etc. In nature, chemicals are neither completely safe nor completely harmful. The safe or harmful effect of a toxicant mainly depends on the dose and duration of exposure of the substances. In addition, the nature and severity of the toxicants also depend on their origin. Naturally occuring toxicants include substances from the animal, plant and mineral sources. A number of microorganisms produce antibiotics some of which cause a toxic impact. The vast majority of toxic substances found in the environment are due to the effluents from industries as well as due to synthetic chemicals of commercial origin.

The toxicants cause physiological, biochemical and histopathological alterations in affected organisms. The impact of the environmental toxicants on organisms can be categorized into two major classes, namely, local effects and systemic effects. Some chemicals cause mild local effects such as irritation at the time of contact or for a shorter duration. On the other hand, severe local action results in the destruction of tissues with which they come in contact. The systemic effects of the toxicants are brought out by the absorption of the toxicants into the blood of the organism followed by distribution to the site of action.

In an environment, the toxicants can occur in different forms which determine their bioavailability. For example, the dissolved chemicals in water are readily available to the organisms. Generally, the toxicants are accumulated in the tissues of organisms (bioaccumulation) and metabolized (biotransformation). As a result, they can either be persistent or converted into other forms. In this way, the fate of a toxicant in the environment depends on the interactions between biotic and abiotic conditions of the environment. The toxicokinetic processes of the toxicant are mainly controlled by the physico-chemical nature of the toxicant, the physico-chemical and biological characteristics of the ecosystem and the sources and quantity of the toxicants present in the environment.

From the human health point of view, environmental toxicants are of various kinds including carcinogens, teratogens, mutagens, cardiotoxicants, immunotoxicants, hepatotoxicants, neurotoxicants, haemotoxicants, reproductive toxicants, etc.

BEHAVIOUR OF TOXICANTS IN THE ENVIRONMENT

In the environment, immission (exposure) and emission (discharge) of a substance occurs at totally different places due to physical or biological factors. Most of the substances are distributed heterogeneously throughout the environment and accumulate elsewhere after dilution. The distribution of substances in the environmental compartments (soil, water and air) mainly depend on the properties of the toxicants such as volatility, water solubility and lipophilicity. These adsorption processes may lead to an increase or decrease in exposure of the toxicants to the organisms. The emitted toxic substances may undergo alterations in the environment (degradation or decomposition, biodegradation, photodegradation, etc.) due to physico-chemical and biological factors. The substances that are non-degradable persist in the environment. The persistence of a toxicant determines its harmfulness to the environment. For example, a highly toxic substance that is easily degraded would be less harmful than a substance that is less toxic but persistent.

The transport of toxic substances lead to the exposure and uptake of the substance by the organisms. The total quantity of the substance taken up by the organism is the bioavailable fraction of a substance. Most of the xenobiotics are also distributed heterogeneously over various organisms and accumulated in their body. The extent of accumulation of xenobiotics in an organism depends on three important factors such as position in the food chain, body size and physiological adaptations. In a food chain, there is loss of biomass but the toxic substances concentrate more and more and reach high levels in top predators. Small-sized organisms take up the toxic substances rapidly due to their large body surface area and high metabolism. On the other hand, large animals take up smaller quantity of substances and have slow metabolism. Therefore, the persistent substances easily accumulate in large animals. Moreover, the physiological effects such as factors related to the uptake of the substance, the internal storage, the processing mechanisms and excretion routes determine the uptake of the substances in an organism.

EFFECTS OF XENOBIOTICS

Though the problems of pollution are generally evident at population and ecosystem levels, they begin at the level of individual organisms. Different types of pollution and the effect of pollutants at various levels in the environment are schematically given below.

Environmental contaminants

Hydrosphere

Lithosphere   ⟶   Bioavailability

Atmosphere

Effects on organisms

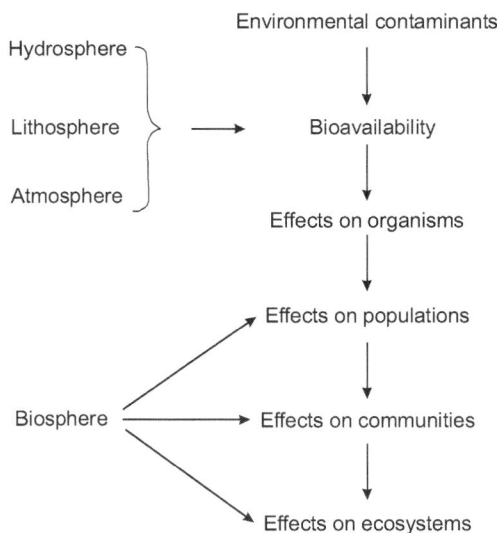

Effects on populations

Biosphere   ⟶   Effects on communities

Effects on ecosystems

Effects on Populations

The sensitivity of organisms to a substance greatly varies within a population due to differences in age, size, nutritional status, sex, etc. A population may have both sensitive and less sensitive organisms depending on the circumstances. To estimate the effect of xenobiotics on population, life cycle toxicity tests are to be carried out by rearing the young ones as test animals at different concentrations of the toxicant in order to derive the differences in survival, growth and reproduction. The other parameters like carrying capacity (maximum number of organisms which can survive in a certain area) and intrinsic biomass turnover (the ratio between the rate of production of organisms in a population and the total biomass) can be analysed.

At population level, the toxicants exert two types of effects, namely, effects on reproduction and effects on survival. Populations can also be classified as food-regulated populations (can withstand temporary food shortage) and predator-regulated populations (with surplus of food). Therefore, the food-regulated populations are more susceptible to those toxicants, which influence the survival of the organisms, whereas the predator-regulated populations are mainly susceptible to those substances, which affect the reproduction of the individual. Hence the organisms at the base of the food chain are very alert to the effects of substances on reproduction. The carnivores at the top of the food chain are food-regulated (suceptible to the substances which reduce the lifetime).

Effects at Community Level

The effects of toxicants on communities are generally measured through structural characteristics, namely, species diversity, i.e., the number of species, the distribution of number of organisms among species and the comparison to a standard condition. The evaluation can be further improved by classifying the individuals of a community into functional groups rather than counting each species.

The Abundance Biomass Comparison (ABC) method is a recent one by which community curves are constructed by plotting various species in the order of density on the x-axis and total number and biomass cumulatively on the y-axis. In an unpolluted ecosystem, the abundance curve lies beneath the biomass curve whereas it lies above the biomass curve in a polluted ecosystem as shown in Figure 20.1. Therefore, it is clear that the community shifts towards large number of small species at the event of pollution load. This is because pollution inhibits the normal development of a community and continues to have small and rapidly growing species.

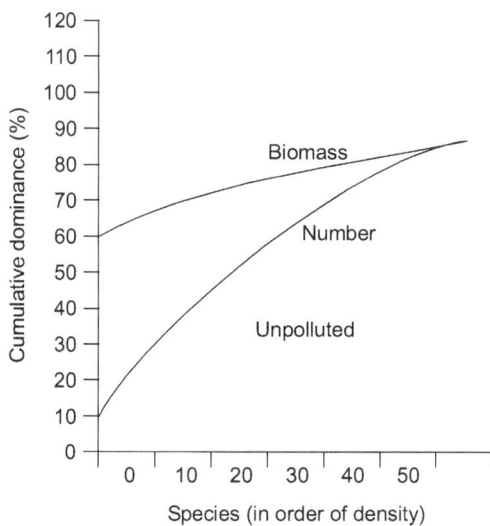

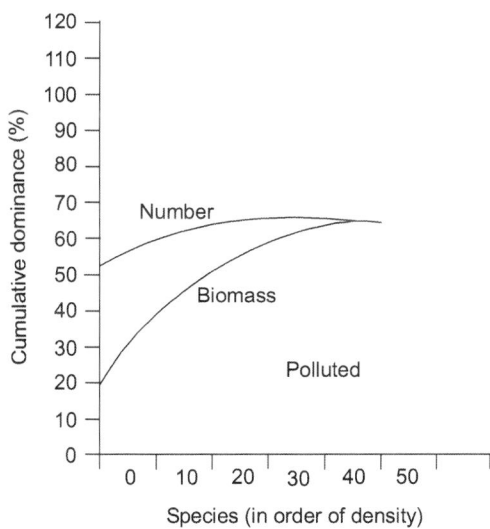

Figure 20.1 Dominance curves for the communities of unpolluted and polluted ecosystems

Effects at Ecosystem Level

In general, the effects of toxic substances at ecosystem level have been evaluated by using model ecosystems. Studies on model ecosystems reveal that they are less sensitive to the toxicants than the individual species. It is also found that the toxic burden disappears rapidly in complicated ecosystems due to degradation or biological unavailability.

Table 20.1 Broad categories of ecotoxicity tests and their uses

Tests	Uses
Single species test	To test on single organism.
Multispecies test	Act as model ecosystems; to test the impact of xenobiotics on a community.
Field ecosystems	To test the impact of chemicals on more representative and complex ecosystems.
Environmental distribution	To test the concentration of toxicants in the environment.
Degradation	To test the persistence of toxic chemicals in the environment, their rates of degradation of breakdown products.
BOD	To test the potential of the substance to cause oxygen depletion in water.
COD	To test the required amount of oxygen for the oxidation of the chemical.
Bioaccumulation	To test the uptake of xenobiotics by organisms.
Bioconcentration	To test the uptake and retention of substances by organisms from the environment.
Biomagnification	To test the uptake and retention of toxicants in the foods of organisms.

Environmental hazard and risk assessment The toxic substances not only cause harmful effects on humans but also bring indirect threatening to his well-being or to his fellow creatures, animals or plants. The environmental hazard and risk assessment are almost identical to that of toxicity tests. But, the environmental hazards are not restricted only to the test organisms. Here the test organisms serve as an indicator species for other similar organisms which might also be affected by the xenobiotics. Again, the mortality of these organisms which play an important role in the food chain could adversely affect the other organisms. The investigation of the effects of poisonous chemicals on the environment is generally referred to as **ecotoxicity testing**. These tests not only examine the effects of substances on organisms but also the fate of chemicals in the environment, their distribution and their effects on other chemical toxicants. The broad categories of ecotoxicity tests and their uses are given in Table 20.1.

REVIEW QUESTIONS

1. Define environmental toxicology and specify its importance.
2. Give an account of the environmental pollution and pollutants.
3. Write a brief account of the environmental contaminants which affect the biota.
4. Explain the various kinds of environmental toxicants and their behaviour in the environment.

5. Give a brief account of the effect of environmental contaminants on the biosphere.

6. Describe the impact of xenobiotics on populations.

7. How do xenobiotics affect communities?

21

PERSISTENCE AND RESIDUAL NATURE OF TOXICANTS

On toxicological point of view, the biosphere includes four environmental compartments, namely, air, water, soil and biological compartments. **Environmental chemistry** deals with the studies on occurrence of chemicals in the environment, their distribution in the environmental compartments and physico-chemical nature and behaviour of substances. The main aim of ecotoxicology is to recognize, predict and analyse the effects of potentially toxic substances. The substances which are directly emitted from the sources into the environment are called **primary pollutants** whereas the **secondary pollutants** are formed in the environment under the influence of the primary pollutants. The concentration of substances in the environment are expressed in different units as in Table 21.1.

Table 21.1 Units of concentration of substances in the environment

Compartment	Units	Abbreviation
Air (gases)	$\mu g/m^3$	ppbv
	$\mu l/m^3$	ppmv
	$\mu l/l$	ppm

(*Contd.*)

Table 21.1   (Continued)

Compartment	Units	Abbreviation
Water	μg/l	ppb
	μg/l	ppm
	μmole/l	μM
Soil	μg/kg	ppb
˙	mg/kg	ppm
	μg/g	ppm
	μmole/g	μM

FATE OF SUBSTANCES IN THE ENVIRONMENT

The fate of substances in the environment is mainly due to **emission** (discharge) and **immission** (exposure) which depend on a number of factors such as distribution, adsorption, transformation and bioavailability.

Distribution

The physical factors such as water, air, etc. as well as the biological factors like migration of organisms would cause emission (discharge) and immission (exposure) of substances at totally different localities. Though distribution results in the dilution of substances, the toxicants will not disappear but they tend to spread heterogeneously in one part of the environment and could accumulate again in some other places.

Adsorption

It is a surface phenomenon by which the substances accumulate at the surface of various environmental compartments. This process mainly depends on volatility, hydrophilicity and lipophilicity of substances. Adsorption may increase or reduce the exposure of substances in the environment. For example, higher concentrations of substances in the sediments of water bodies would lead to lesser exposure of the pelagic organisms but increased exposure of benthos.

Transformation

During transformation, the substances may change in its chemical form due to either physico-chemical factors or biological factors. These alterations are called **degradation** or **decomposition**. Alteration caused by light is called **photodegradation** and by microorganisms is called **biodegradation**. The substances, which do not decompose or degrade very slowly in the environment, are said to be persistent. **Persistence** is one of the most important factors, which determines the toxicity of substances in the environment. A number of processes such as dispersion, physical changes and chemical changes take place during the transport of substances from the source to the organisms. The quantity of substances discharged was expressed in weight per unit time. After the exposure, a part of the substances become dissolved in water or volatilized in air, the concentration of which is called **exposure concentration.** Another part of the substances is deposited into subterranean water and surfaces of plants and animals. This process is referred to as deposition, which is expressed in unit weight/surface area/unit time. The risk analysis of toxic substances therefore requires data on emission, deposition, behaviour in various environmental compartments and their toxicity.

Bioavailability

The transport process ultimately leads to exposure and uptake of the substances by the organisms. However, only a part of the total concentration of the substance is actually taken up by the organism (**bioavailability**). In cases where the total amount of substances remain constant, the bioavailability changes because of a balance between the bound and suitable fractions of the substances in the environment.

ACCUMULATION OF SUBSTANCES IN ORGANISMS

The concentration of a substance in an organism is determined by the balance between uptake, conversion and excretion of the substances. The residue depends on the nature of substances and the organism itself. Therefore, the accumulation of substances represent the quantity present in an organism (body burden). The time during which a substance remains in the body of organisms is called **half-life** of the toxic substance (t ½). Lipophilic substances such as PCBs and some chlorinated hydrocarbons have long half-lives, so they accumulate readily in organisms. Therefore the main criteria for accumulation is frequent exposure to the substances which have high (t ½) values.

The following are three important factors that affect the extend of accumulation of substances in the body of organisms:

1. Position in the Food Chain

In general, the number of organisms and their biomass decrease towards the higher levels in the food chain. During the transition between the tropic levels, there is loss of biomass but not of persistency of substances. Therefore, the substances concentrate more and more, culminating at the highest level in top predators (**biomagnification**). For example, DDT residues in the water of a marsh habitat is about 0.00005 mg/ml whereas in fishes it is 0.3 mg/g and in birds 2.5 mg/g.

2. Body Size

Generally, large organisms consume less food per unit body weight and possess lower metabolic rate than small organisms. This indicates that the oxygen consumption is not directly proportional to the size of the organism. This means that the oxygen uptake per unit weight of the body decreases with the increasing weight (**allometric relationship**) which is due to the changing ratio between the surface area and volume of the animals. In other words, the volume of an animal is directly proportional to its weight but the surface area increases with its weight to the power of 2/3. In organisms, the uptake of substances always occurs via the surface while the processing of substances is proportional to the metabolism. Thus, as small organisms have relatively large body surface area per unit weight and high metabolic rate, they will take up substances rapidly. At the same time, it is also true that the concentration of the substances in the body of these organisms fall very rapidly. On the other hand, the larger organisms possess relatively low metabolic rate and will take up and eliminate the substances very slowly. Therefore, the persistent substances accumulate easily in the body of large animals. It is noteworthy that the organisms in higher tropic levels in food chains are larger.

3. Physiological Variables

The uptake and the residual level of toxic substances in organisms mainly depend on physiological variables such as uptake of the substances by the organisms, the possibilities of internal storage, processing mechanisms and excretion routes. As an example, the hepatopancreas of woodlice concentrates large quantity of copper as well as other heavy metals like lead, cadmium, etc. The storage capacity of organisms mainly depend on the fat content of the organisms as the fat has affinity to lipophilic substances like organochlorine compounds. The processing and excretion mechanisms vary from animal to animal and also in various substances. Many aromatic substances are converted into their metabolites through MFO system and are excreted. Poor metabolizing capacity of substances by the organisms would lead to the accumulation of high levels of substances in their body. For example, the fish-eating birds are unable to metabolize the organochlorine compounds.

FATE OF PESTICIDE RESIDUES

While applying the pesticides over wide areas of agricultural lands, soil, water bodies, human settlements and forests, a reasonable quantity of the chemical reaches the environment, thus forming a major source of environmental contamination. Spraying of the pesticides cause drifting in air currents from the point source so that the pesticides may settle on land and water very close to the point of application or may reach the atmosphere along with wind, thus carried to far away places. If they get degraded in the environment into harmless substances, then there would be no environmental degradation. But major portion of applied pesticides are transported into the environmental compartments, thereby affecting the ecosystems and communities.

The toxicity, persistence and specificity are the characteristics of pesticides. Persistence is the length of time that the chemicals remain in the environment or in the organisms without being broken down. While biodegradable substances decompose quickly, certain substances possess a high degree of resistance to decompose. These substances persist in the environment, become accumulated and biomagnified and exert harmful toxic impact in organisms at higher trophic levels of the food chain. Many pesticides are highly persistent in a single ecosystem or may become persistent when the residues move from one ecosystem to another. For example, endosulphan is a highly toxic pesticide, which possesses the property of longer persistence and slower degradation. Comphenchlor and DDT are extremely stable compounds and the former persists as residue for months and the latter for years. In general, organochloride pesticides in aquatic ecosystems exhibit greatest persistence, organophosphates and carbamates exhibit intermediate persistence and herbicides, the least persistence. However, a few like malathion are very unstable and degrade in few hours or in few days. Based on the duration of existence in the environment, the pesticides can be categorized into five major groups as in Table 21.2.

Though the persistence of chemicals is undesirable quality, it is required to control animal parasites and soil-borne diseases. But long-term persistence would lead to very damaging impact on organisms.

To cope up with the requirements, different pesticides in different combinations and formulations are being used.

This heavy usage of pesticides is the main reason for the occurrence of large quantity of pesticides in the biota and the environment. Thus a contaminated ecosystem may contain different pesticides at varying concentrations. As a result, the toxicity of each pesticide is influenced and modified by the other pesticides (**pesticide interaction**). The interaction would occur not only between the chemical pesticides but also with abiotic and biotic factor of the ecosystem. Studies on interaction of various drugs during absorption, distribution, accumulation and excretion have been well established but only very little information is available with regard to the pesticides. Thus there is limited knowledge of pesticide interactions on non-target organisms as the studies are restricted only to insect and mammalian species.

Table 21.2   Categories of pesticides based on duration of persistence

Group	Examples	Duration of persistence
I	Organochlorine pesticides	For years and more than 18 months
II	Derivatives of diazine, urea, etc.	About 18 months
III	Derivatives of various acids	About 12 months
IV	Derivatives of toluidines,nitryl, etc.	About 6 months
V	Pyrethroids and carbamates	About 3 months or below 3 months

Pesticide Residues in Soil

Direct application to the soil, incorporation to the topsoil and application onto crop plants are the major routes through which the pesticide residues accumulate in the soil. When the residues come in contact with the soil, they are absorbed by the clay and organic matter of the soil. Water-soluble residues become leached into the soil along with percolating water and contaminate the groundwaters. The pesticides are found to degrade very slowly in underground waters. In soil, the residues become chemically degraded and engulfed by soil organisms. They may also be volatilized and may escape into the atmosphere. The decomposition rates of pesticides vary and they persist in the soil for different periods. For example, synthetic pyrethroids and carbamates decompose within three months. Certain herbicides are stable for 18 months. Toxaphene, an organochlorine pesticide has a half-life of 10 years. In the soil, the pesticide residues exert phytotoxic effect in addition to toxic impact to consumers. They also reduce the activity of microorganisms, so the degradation of pesticides in the soil is reduced.

Pesticide Residues in Water

Surface waters are contaminating by the pesticide residues through run-off waters from agricultural lands, which are subjected to pesticide application. The residues adhered to the soil particles may be carried into water courses by soil erosion. In water, traces of pesticides may accumulate progressively in different steps of food chain. When pesticides in water are consumed by planktons, they accumulate in their body tissues (bioconcentration). When the fishes feed on planktons, the pesticide residues enter the food chain and accumulate in many fold increased concentrations. According to one estimate, a very low concentration of DDT in water gets magnified 7000 times in bivalves within 40 days.

Pesticide Residues in Air

Pesticides get entry into the atmosphere through drifting and volatilization during application and escape from pesticide manufacturing plants. The suspended particles in the air absorb the pesticide residues, which are carried to large distances by the wind. The wind also acts as a carrier of volatile pesticides. Thus they occur either as particulate deposits or as gaseous phase. Pesticide level in air is usually higher during summer than during winter. In air, chlorinated hydrocarbons and organophosphates do not degrade so that they persist for many years. Thus these are the most important pesticides which contaminate the atmosphere.

Pesticide Residues in Food

The food materials get contaminated with pesticide residues due to growing of crops in soil containing persistent pesticide residues. Many pesticides have affinity to lipids so that they accumulate in animal tissues, thus contaminating the animal products with residues. DDT and BHC are found to be present in many food materials in more than the tolerance limits prescribed by WHO/FAO. In this way, major food crops are shown to contain residues of endrin, heptachlor and endosulphan. Equally high quantities of residues are also detected in meat and poultry. Many pesticide residues are also reported to be present in breast milk, cow's milk, butter and ghee.

Pesticide Formulation and Toxicity

The pesticides are manufactured as emulsifiable concentrates (EC), wettable powders (WP), granules, dusts, aerosols, etc. in order to disperse the substances on target areas. Different types of carriers such as emulsifiers, diluents, etc. are also used for active dispersion of pesticides. These carriers are often toxic or may interact with the pesticides resulting in more toxicity. In general, emulsifiable substances are more toxic than technical grade compounds whereas dust and powder formulations are less toxic. Moreover, the isomers of a compound are mobile in the soil so that the aquatic organisms may be exposed only to the isomers and not to the pure compounds. In addition, the degradation products of a compound may be more toxic to the organisms than the parent compounds.

Toxicity of Pesticide Residues in Organisms

Pesticides are extensively used to eliminate unwanted pests in order to boost over agricultural production. The organochlorine and organophosphate pesticides are highly stable synthetic organics and possess broad spectrum of action on the nervous system of insects. However, most of the pesticides are highly toxic to many non-target organisms. In addition to the toxicity of individual pesticides on non-target organisms, the introduction of two or more chemicals into biological systems could produce cumulative action (synergism). For example, malathion, a safer organophosphate pesticide causes toxicity in vertebrates in combination with small quantity of EPN or phenthoate, another organophosphate pesticide. Organochlorine compounds are in general found to be more toxic to fishes when compared to the organophosphate pesticides. Among the organochlorine pesticides, endrin and cyclodienes are highly toxic, DDT has greater chronic toxicity. Among the photoisomerization products of cyclodienes, photoaldrin and photodieldrin are more toxic than their normal isomers. In fishes, organophosphates exhibit negligible chronic toxicity whereas the carbamates are moderately toxic.

Pyrethroids possess high insecticidal activity, though they are not persistent in the environment. But they exert high acute toxicity to fishes. Fenvalerate and permethrin are more toxic than many organochlorine pesticides. Though few herbicides exhibit chronic toxicity in fishes, they show only low acute toxicity. In general, different life stages of fishes differ in their sensitivity to different pesticides. The pesticides are shown to affect fertility and immune system and are found to be teratogenic, carcinogenic and mutagenic in experimental animals and mammals.

Residue Standards

The highest dose at which there are no-observable-effects in most test organisms is called **no-observable adverse effect level** (NOAEL). The quantity of a chemical that can be consumed everyday over the entire life of an organism without adverse effect is **acceptable daily intake** (ADI), which is expressed in mg of the chemical in the food/kg of the body weight/day. The maximum concentration of a residue, which is permitted in the food to be consumed, is the **maximum residue limit** (MRL). MRLs will vary from country to country because of differences in agricultural practices, climatic conditions, pests, crops and so on. ADIs and MRLs of pesticide residues are recommended by FAO/WHO.

FATE OF HEAVY METALS

Metals having an atomic number more than 20 except alkaline metals and alkaline earths are heavy metals. Copper, cadmium, cobalt, silicon, zinc, nickel, lead, arsenic, mercury, etc. from industrial effluents, burning of fossil fuels, domestic sewage and land run-off enter the environmental compartments and exert their impact on living organisms.

Most of the aquatic ecosystems receive excessive heavy metals as by-products of industrial processes and acid-mine drainage residues. They are also released from natural sources like withering of rocks and volcanic activity. They are neither degraded nor metabolized so that they are highly persistent in nature. They also possess property of accumulation like pesticides over a long period of time along the food chain. For example, both organic and inorganic forms of mercury are highly poisonous. When mercury compounds are discharged into the water bodies, they are metabolically converted into methyl mercury compounds in the bottom muds by microbial action. Methyl mercury is highly persistent and accumulates in the food chain. Its half-life in man is found to be 70 days. As it is highly lipophilic, it is readily absorbed by the organisms and stored in fatty tissues. Similarly, fluoride universally occurs in soil, water, atmosphere and plant and animal tissues. It is not absorbed in the blood of man but has an affinity for calcium so that it accumulates in the bones of human beings and cattle. Higher concentration of fluorides in plant tissues cause fluorosis.

FATE OF TOXICANTS IN THE ATMOSPHERE

At present, the entire biosphere faces increased demands for energy, air, water, and land. Even though air carries a number of pollutants and releases them into the atmosphere, the carrying capacity of the air is limited. In some tropical areas, the atmospheric pollution occur at high levels, requiring necessary control measures. The following are the toxic pollutants present in the atmosphere.

Carbon Compounds

These include carbon dioxide released by complete combustion of fossil fuels and carbon monoxide released from automobile exhausts. Carbon dioxide is confined to troposphere where it absorbs the heat generated from the earth. As a result, the temperature of earth's atmosphere is increased at global level (greenhouse effect). It is predicted that the temperature of the earth could increase by 1.5 to 4.5°C by the year 2050. The major source of CO is the exhaust products from motor vehicles, in addition to all other combustion sources. CO can kill organisms at concentrations over 1000 ppm and the lesser concentrations would lead to poison symptoms. It reduces the oxyphoric capacity of the blood in higher animals.

Sulphur Compounds

These include sulphur dioxide, hydrogen sulphide and sulphuric acid. The complex mixture of sulphur dioxide, sulphuric acid and inorganic and organic dust particles constitutes an important component of the air pollution. In the atmosphere, the major sulphur-containing substances are carbonyl sulphide, carbon disulphide, dimethyl sulphide, hydrogen sulphide and sulphate. Sulphur dioxide is emitted to the atomosphere from burning of fossil fuels by industries. Hydrogen sulphide is a colourless toxic gas and is found to be fatal at concentrations of 700–900 ppm through respiratory system. Its major sources are industries, which use sulphur-containing fuels.

Nitrogen Oxides

The atmosphere contains nitrogen oxides such as nitrous oxide, nitric oxide, nitrogen dioxide, etc. of which nitrogen monoxide and nitrogen dioxide occur significantly in the atmosphere threatening to the public health. Combustion in power plants, industries and domestic practices, automobile emission, etc. are the main sources of nitrogen oxide in the atmosphere. Natural emissions from soil and lightning as well as anthropogenic emissions also act as sources. The nitric acid is readily converted into nitrate particles in the atmosphere, which involve in photochemical processes like formation and depletion of ozone. The oxidized nitrogen compounds from the atmosphere are incorporated into rain or snow or directly deposited on earth's surface. Sulphuric acid and nitric acid play a major role in acid rains which cause increased acidity in the soil, threaten life of organisms, bring corrosion, destroy forests, reduce agricultural productivity and so on.

Fluoride Compounds

This may occur either in gaseous or particulate state in the atmosphere. The main sources of these compounds are the industries involved in the manufacture of fertilizers, aluminum, fluorinated hydrocarbons, plastics, ceramics, etc.

Hydrocarbons

These are organic compounds emitted into the atmosphere by the evaporation of gasoline, benzene, benzopyrene, methanol, etc.

Metals

Though zinc is a natural constituent of air, the industries involving in copper, lead and steel refining are the main sources of atmospheric zinc. In the atmosphere, zinc occurs in the form of white zinc oxide fumes. Lead is a naturally occurring metal found in all parts of the environment. The main sources of lead come from human activities such as mining, manufacturing and burning of fossil fuels.

The major source of lead into the environment is the use of lead organic compounds as motor vehicle fuel additives. Tetraethyl lead compounds added to gasoline are emitted into the atmosphere as volatile lead bromides and chlorides. Dispersion of mercury in the atmosphere occurs through waste incineration and manufacturing of fungicides, papers, paints and cosmetics. Mercury combines with other elements like chlorine, sulphur, oxygen, etc. to form inorganic compounds. The main source of cadmium in the atmosphere is industries which are involved in processing of cadmium-containing materials and in refining of copper, lead and zinc. It is also released into the atmosphere during the production of pesticides and phosphate fertilizers. Cadmium occurs as oxides, chlorides, sulphides and sulphates. The other metal residues in the atmosphere and their sources are given in Table 21.3.

Table 21.3   Toxic metals and their sources

Metal	Sources
Arsenic	By-product of metal refining operations
Chromium	From steel industries, tannery units and chemical industries
Nickel	From chemical petroleum and metal production industries

Ozone

Ozone is found only in minute quantities in the atmosphere but it is highly concentrated in the stratosphere. It is produced by the oxygen-containing substances such as sulphur dioxide, nitrogen oxide and aldehydes on absorbing ultraviolet rays. It absorbs all radiations below 3000 Å which are harmful to the organisms. The increased use of chlorofluorocarbons is the main cause for the depletion of ozone layer on earth's atmosphere. The chlorofluorocarbons are emitted by aerosols and escape into the stratosphere where they react with ozone resulting in warming of the earth.

Photochemical Products

The aromatic compounds which arise photochemically in the atmosphere are benzopyrene, peroxyacylnitrate (PAN) and peroxybenzylnitrate. The photochemical smog includes ozone, nitrogen compounds, hydrogen peroxide, organic peroxides, PAN, etc. which are produced photochemically in the atmosphere.

Aerosols (Particulate Matter)

These are the materials existing in the atmosphere as liquids or solids. The primary aerosols are directly emitted into the atmosphere from the sources (wind or smoke) and the secondary aerosols are produced by the chemical interactions of the gases. The main sources of particulate matter include the fuel combustion and industrial operations. Thus mining, polishing, textile, pesticide and fertilizer industries as well as the industries which handle loading and transfer operations are the main sources of particulate matter in the atmosphere. In addition, the non-industrial sources include roadway dusts, agricultural and construction operations, and transportation sources such as vehicle exhaust, particles smaller than 10 milli-micron in diameter such as soot, lead particles, asbestos, volcanic emission, dust particles, pesticides and also biological particulates like bacteria, spores, pollen, etc. cause health disorders in man.

REVIEW QUESTIONS

1. What are the primary and secondary pollutants?
2. Explain the factors that determine the fate of toxicants in the environment.
3. Describe the various factors that affect the extent of accumulation of toxicants in the body of organisms.
4. What is allometric relationship?
5. Write an essay on pesticide residues.
6. Give an account of pesticide residues in soil, water, air and food.
7. Write an essay on fate of heavy metals in the environment.
8. Describe the toxic pollutants found in the atmosphere.
9. What are aerosols? Explain types of aerosols.

22

BASICS OF NUTRITIONAL TOXICOLOGY

Nutritional toxicology is the combination of toxicology with nutrition and food science, thus focusing on the toxicological aspects of foodstuff and nutritional habits. It deals with the possible adverse effects of dietary substances on human health. The main objectives of nutritional toxicology include recording and increasing the knowledge and understanding of toxic foodstuff, assessing their risks based on the levels of their occurrence, dietary habits and lifestyles of consumers and furnishing information to the society to prevent toxicological risks of foodstuff.

Potential Toxicants in the Foodstuff

The potential toxicants in the foodstuff can be broadly classified into (i) natural compounds (ii) compounds that enter foodstuff directly or indirectly due to industrial processes and (iii) additives which are deliberately used in food production.

Natural Toxic Compounds

In nature, many food sources such as marine toxins, glycoalkaloids and lectins contain toxic compounds.

The marine toxins are mostly produced by single-cell organisms like dinoflagellates and occur secondarily in animals. Glycoalkaloids like solanine and chaconine are found in vegetables like potatoes and tomatoes. Consumption of these vegetables with high content of glycoalkaloids causes poisoning. The lectins are haemagglutinins which are toxic proteins found in the seed of certain plants. In recent times, these food toxins are shown to cause several diseases in human as well as in cattle and poultry. Table 22.1 gives an account of toxicants found in plants and microorganisms and their impact on human.

Table 22.1   Some naturally occurring toxic plants and microorganisms and their impact on human

Plants/Microorganisms	Impact
Seeds of *L.sativus* and other *Lathyrus* species	Lathyrism (crippling disease followed by paralysis of leg muscles)
Ackee fruit	Hypoglycaemia resulting in coma and death
Brassicae family (cabbage and turnip), oil seeds like rape seed, mustard, etc.	Goitrogen
Bananas, lemons, pineapple, etc.	Cardiac problems
Argemone seed	Epidemic dropsy
Microorganisms	
Salmonella	Vomiting, diarrhoea, increased body temperature
Staphylococcus	Vomiting, diarrhoea, abdominal pain

(Contd.)

Table 22.1 (Continued)

Plants/Microorganisms	Impact
Shigella	Dysentery
Aspergillus flavus	Cancer and cirrhosis
Clavoceps purpurea	Vomiting, diarrhoea, gangrene in fingers and toes, convulsions
Fusarium and *Cladosporium*	Gastro-enteritis, leukopenia, anaemia, necrotic ulcers in skin

Industrial Contaminants

These include heavy metals, a variety of chemicals, plastic components, pesticides and veterinary medicines. These contaminants find their way into the food sources from the environment. The polychlorinated insecticides have been given much concern as relatively high concentrations of these compounds are identified in breast milk. The pesticides and veterinary medicines which are used to eradicate pests and diseases adversely affect the production of foodstuff as the residues find their way to human beings. Processing of foodstuff (mechanical treatment, heating, smoking, etc.) often cause changes in chemical composition due to the addition of compounds like traces of metals. During packaging, contaminants may enter into the packaged food. For example, microbial toxins (endotoxins) and botulinum toxins (exotoxins) contaminate the foodstuff during packaging or food processing. Certain potentially toxic substances like micotoxins and pesticides which are used during storage and transportation of food materials can enter and contaminate the foodstuff. Table 22.2 shows the effect of metals and pesticides found in the foodstuff.

Table 22.2   The effect of metals and pesticides found in various foodstuff

Name of the contaminant	Foodstuff	Effect
Mercury	Seed grains	Nervous poison
Cadmium	Soft drinks and fruit juices	Liver and kidney damage
Lead	Processed foods	Nervous poison
Arsenic	Fruits	Nervous poison
Cobalt	Beer	Nervous poison
Tin	Canned foods	Vomiting
Zinc and copper	Stored foods	Diarrhoea
Pesticide	All foodstuff	Liver, kidney and brain damage

Food Additives

The chemicals that are deliberately added to the foodstuff in order to have some desirable quality and to improve their storage life, stability and flavour are called "**food additives**". These include antioxidants, emulsifiers, flavour and colour imparting chemicals, preservatives, sweeteners, etc.

Colouring agents   Various foods, canned fruits, soft drinks, etc. are added with colouring agents which may be natural or synthetic. The natural colours are obtained from plant pigments and the Table 22.3 gives various pigments and their sources.

The main group of synthetic colours are the azodyes which are shown to be allergic and carcinogenic agents. Some substances like saffron, sandalwood, turmeric, etc. are used as colouring as well as flavouring agents.

Table 22.3 Pigments and sources

Pigment colour	Sources
Yellow, orange and red	Carrot, orange, tomatoes
Green	Leaves of plants
Red, blue and violet	Beetroot, plums, cabbage
Yellow-coloured flavonoid pigments	Leaves and petals of plants

Flavouring agents These may also be natural or artificial. Vanilla, saffron, cocoa, turmeric, pista, etc. are common natural flavouring agents obtained as plant products. Synthetic flavouring agents include monosodium glutamate, guanosine 5'-disodium phosphate, inosine 5'-disodium phosphate, sodium 5'-ribonucleotide, etc.

Preservatives Vinegar, citric acid, etc. are used as preservatives to improve the durability of foodstuff and to prevent their spoilage and food poisoning.

NUTRITION AND TOXICITY

A healthy person is normally less susceptible to the toxicity of chemicals when compared to a person with a poor state of health. Experimentally it is proved that certain dietary deficiencies increase the sensibility of animals to the toxicants. For example, the deficiency of iodine makes the thyroid glands more sensitive to antithyroid compounds. Thiamine deficiency is shown to increase the cardiotoxic effect of certain heavy metals.

Diet influences the constitution of biochemical processes in the body as well as the physiological processes. For example, the enzyme system in the liver controls the biochemical conversion of the foreign substances and so they are called biotransformation enzymes. These enzymes can either activate certain substances into toxic products or can detoxify them. The activity levels of biotransformation enzymes are affected by macronutrients such as proteins, lipids and also by various non-essential compounds found in the foodstuff such as indole coumpounds. The balance between the activating and detoxicating the functions by the biotransformation enzymes, thus determines the availability of toxic metabolites in the body.

REVIEW QUESTIONS

1. Classify the potential toxicants of foodstuff.
2. Give an account of natural toxic compounds.
3. What are the industrial contaminants found in food sources?
4. What are food additives?
5. Write an essay on nutrition and toxicity.

23

LIVER FUNCTION TESTS IN TOXICOLOGY

Liver plays an important role in certain synthetic and degradation processes mediated by hepatic enzyme systems. It participates in digestive process by secreting bile, in enterohepatic circulation and in the metabolism of proteins, carbohydrates and lipids. It is responsible for the biotransformation of foreign substances through oxidation, reduction, hydrolysis or conjugation. As a result, the lipophilic chemicals are converted into hydrophilic substances to be excreted. Because of its metabolic, detoxicating, secretory and excretory functions, the liver is highly susceptible to toxic substances which may interfere with the metabolic processes in the liver or with the secretion of bile resulting in liver disorders.

HEPATOTOXINS

Some xenobiotics, while entering into the liver, cause direct cell injury whereas others undergo metabolic conversion resulting in the formation of toxic by-products. Such substances which cause hepatotoxic effects in animals are called **intrinsic hepatotoxins.** They interfere with the hepatocellular metabolic processes or bile secretion causing lesions characteristic of cytotoxic and cholestatic changes in the liver such as acute or chronic hepatitis, necrosis, steatosis, cholestasis, phospholipidosis, cirrhosis and tumours.

These hepatotoxins are classified into direct and indirect intrinsic hepatatoxins. The **direct hepatotoxins** such as phosphorous, tannic acid, etc. produce primary lesions in the liver resulting in disturbed cell metabolism. The **indirect hepatotoxins** such as steroids, ethionine, etc. interfere with the cell metabolism with the resultant loss of cell integrity. In addition, some xenobiotics may also cause hepatotoxic effects through immunological mechanism. A xenobiotic or one of its metabolites may cause a delayed immunological effect leading to hepatocellular necrosis. The toxic substances may cause either **predictable** or **non-predictable liver lesions.** The predictable impacts include hepatitis caused by methotrexate, necrosis by carbon tetrachloride, jaundice by novobiocin, liver steatosis by tetracycline, etc. In non-predictable reactions, the toxic effects of xenobiotics on organisms cannot be foreseen. These include cholestasis induced by steroids, hepatitis by chlorpromazine, etc.

Chlorpromazine and steroid compounds inhibit bile flow and excretion of organic anions such as bilirubin and bile salts thus inhibiting the active transport of bile acids in the liver **(cholestatic disturbance)**. The liver steatosis is caused by many toxic substances due to an imbalance between synthesis and secretion of triglycerides, increased synthesis of free fatty acids from acetyl-CoA by reduced lipoprotein production, the disturbance in the elimination of triglycerides through interference with the mitochondrial oxidation of fatty acids and with the synthesis of lipoproteins or the inhibition of protein synthesis. Table 23.1 gives an account of various liver disorders caused by different hepatotoxins.

Table 23.1 Disorders caused by hepatotoxins

Liver disorders	Toxins
Cytotoxic lesions such as steatosis, necrosis and apoptosis	Carbon tetrachloride, cytostatics, salicylates, acetaminophen, tetracyclines, tuberculostatics, anti-inflammatory agents, halothane, etc.
Cholestatic lesions (both intra hepatic or hepatocellular and extrahepatic or canalicular)	Chlorpromazine, tolbutamide, oral contraceptives, anabolic steroids, etc.
Hepatitis	Viruses, alcohol and drugs such as isoniazid, sulphonamides, propyl-thiouracil, acetaminophen, halothane, rifampicin, methyl dopa, methotrexate, etc.
Liver necrosis	Tetrachloroethane, trinitrotoluene, dinitrobenzene, mixtures of chlorinated biphenyls, naphthalenes, etc.
Phospholipidosis and fibrosis	Amphophilic cationic substances
Cirrhosis	Methotrexate, methyl dopa, etc.
Hepatovascular lesions	Oral contraceptives, dimethyl-nitrosamine, pyrrolizidine alkaloids, etc.
Tumours	Hepatocarcinogens (Natural–aflatoxin B, pyrrolizidine alkaloids, cycasin, safrole, etc. Synthetic–dimethylnitrosamine, diethylnitrosomine, DDT, PCBs, CCl_4, chloroform, vinyl chloride, etc.)
Granulomatous lesions	Sulphonamides, methotrexate, halothane, phenylbetazone, etc.

LIVER FUNCTION TESTS

The liver function tests are used to study the liver and biliary tract disorders in affected organisms including human. The liver disorders can be identified by analysing serum bilirubin, serum cholesterol and plasma proteins, namely, albumins and globulins. However, many liver function tests are mainly based on the estimation of the activity levels of serum enzymes like glutamate oxaloacetate transaminase (SGOT) or aspartate aminotransferase (AST), glutamate pyruvate transaminase (SGPT) or alanine aminotransferase (ALT) and serum alkaline phosphatase (SAP).

Albumin and Globulins

Albumin is the most abundant plasma protein to which fatty acids become bound. From the albumin, the fatty acids are released and are taken up by the tissues where they serve as gluconeogenic substrate. It maintains the osmotic balance inside and outside of the blood vessel through osmotic and hydrostatic pressure. It acts as the transport molecule for substances like calcium, bile salts and hormones. The albumin level changes only slowly in liver disorders. Cirrhosis, a liver disorder exhibits hypoalbuminaemia due to reduced synthesis and turnover of albumin from amino acids. In chronic hepatocellular disorders, hyperglobulinaemia occurs. Globulins are large molecules of plasma proteins. Among them, α and β globulins act as transport vehicles for hormones, cholesterol, lipids and vitamins whereas γ globulins are anitibodies involved in immune reactions. In chronic liver diseases, the serum albumin level falls but the serum globulin level increases.

Transaminases

Transamination is a reversible reaction in which one amino acid transfers its amino group to a keto acid resulting in the formation of corresponding amino acid and α-keto glutarate. Glutamate and aspartate are the major amino acids which undergo transamination to produce citric acid cycle intermediates, namely, α-ketoglutarate and oxaloacetate respectively. The transamination reactions are catalysed by transaminases or aminotransferases. The transaminases play a key role in serving as a link between the metabolism of carbohydrates and proteins in the interconversion of ketoglutarate, pyruvate and oxloacetate on one hand and alanine, aspartate and glutamate on the other hand. Among the serum transminases, GOT and GPT are the two most important enzymes which mostly use glutamate or α-ketoglutarate. GOT is a mitochondrial enzyme found in heart, liver, skeletal muscle and kidneys. On the other hand, GPT is a cytosolic enzyme and more specific for the liver. Therefore, the liver injury in animals can be tested based on the activity levels of these enzymes in the serum. Any damage to the above organs would cause the release of these enzymes from the cells of the organs so that the activity level of these enzymes would be elevated in the serum.

The reason for the higher activity levels of these enzymes could be due to increased concentration of amino acids most probably, aspartate and alanine. The xenobiotics would stimulate the break down of carbohydrates and proteins leading to accumulation of amino acids which are utilized to cope up the stress condition through increased transamination reaction. Thus transamination reactions via transaminases provide major routes for both synthesis and degradation of amino acids especially glutamate, aspartate and alanine. The increased levels of these enzymes indicate acute heart failure and pancreatitis, muscular dystrophy, leukaemia, hepatocellular necrosis and haemolytic anaemia.

Serum Alkaline Phosphatase

Serum alkaline phosphatase (SAP) is not an organ-specific enzyme and is found in many tissues. In the liver, it is closely associated with lipid membrane and interferes with bile flow. The increased SAP activity in the serum would indicate hepatocellular necrosis.

REVIEW QUESTIONS

1. Explain the hepatotoxins with suitable examples.
2. Name the various liver function tests and mention their uses.
3. Explain the major functions of albumin and globulins.
4. Write an essay on transaminases.
5. What is the significance of serum alkaline phosphatase and γ-glutamyl transpeptidase?

24

ANTIDOTAL PROCEDURES (TREATMENT OF INTOXICATION)

The pathological symptoms noticed in affected organisms are collectively called as **intoxication**. The intoxication is mainly due to physico-chemical reactions that occur at the site of contact between the toxicants and the body of the organisms. When the toxic chemical is absorbed by the body and transported to various organs, it affects and alters the functions of those organs fully or partially. This is **systemic intoxication** which is of two major types, namely, **acute intoxication** and **chronic intoxication.**

ACUTE INTOXICATION

It is due to the exposure of a patient to a single but relatively large quantity of a toxicant for shorter exposure duration. The symptoms of acute intoxication are usually severe and appear soon after the exposure so that the patient is to be subjected to immediate medical attention. The occurrence of acute toxicity varies widely and has a great practical significance. The acute intoxication may either be **accidental (unintentional)** or **deliberate intoxication**. The accidental intoxication occurs unintentionally due to eating or drinking of a substance which is not meant to be consumed or due to consumption of chemicals. Some important groups of such toxicants are shown in Table 24.1. The deliberate intoxication is the administration of toxic substance intentionally by the patient himself. It may be an **attempted homicide** (administration intentionally by a third party) or **attempted suicide** (deliberate auto-intoxication). The problem of deliberate intoxication is often characterized by **combined intoxication,** as the patient takes not only an overdose of toxic chemicals but alcohol too.

Table 24.1   Groups of toxicants which affect humans

Toxic substances	Affected individuals (%)
Cleaning substances	10.4
Analgesics	10.0
Cosmetics	8.3
Plants	6.1
Drugs	5.7
Pesticides	3.8
Hydrocarbons	3.5
Sedatives	3.2
Toxic chemicals	2.9
Alcohol	2.7
Food poisoning	2.5

CHRONIC INTOXICATION

Chronic intoxication refers to a pathological condition due to repeated exposure to a potentially toxic substance at higher concentrations. It is largely associated with occupational exposure or with environmental pollution in which the pathophysiological mechanisms are similar. But the level of occupational exposure to the toxicants is higher than the level in the environmental pollution. Table 24.2 gives a few examples of occupational poisoning

Table 24.2 Toxic chemicals at work and their pathological effects

Toxicants	Pathological effects
Irritant gases—chlorine, metal vapours, etc.	Pulmonary oedema, pneumonia, fibrosis and chronic bronchitis.
Heavy metals—chromium, cadmium, mercury, etc.	Kidney damage and neuropathic effects.
Halogenated hydrocarbons—aliphatic and aromatic compounds	Liver damage and dermatotoxic effects.
Metals—lead, arsenic, nitro compounds, etc.	Blood abnormalities including anaemia and methaemoglobinaemia.
Organophosphate pesticides	Neurotoxic effects.
Solvents, detergents, formaldehyde, vinyl chloride, etc.	Dermatotoxic effects.
Cobalt compounds, carbon disulphide, nitroglycerine, etc.	Cardiovascular diseases.

ANTIDOTAL PROCEDURE

The toxic effect is produced only when the toxicants are absorbed and transported to the receptors or target sites in the body of organisms. The magnitude of toxic effect is determined by the concentration of the toxic chemicals at target sites. But the concentration of toxic substances at the site of action depends on the rate of absorption, rate of transport to the site of action and rate of detoxication and elimination. Therefore, the toxicity of the xenobiotics can be reduced or terminated by applying the following aspects:

i. *Knowledge about the properties of toxic substances*   A sound knowledge of toxicological properties, kinetics and biotransformation processes of the toxicants is required to give appropriate relief to the poisoned person.

ii. *Prevention or reduction of absorption*   The immediate removal of a toxic substance from the intestine can reduce its concentration at the site of action. This could be achieved by inducing vomiting (**emesis**) either through reflex vomiting or using emetic agents like apomorphine. The gastrointestinal content can be siphoned out along with a solvent or a liquid through a tube inserted via the mouth into the stomach. This is actually rinsing or washing of the stomach and is continued until clear fluid is drawn out (**gastric lavage**). Effective adsorbents like medicinal activated charcoal can be used to adsorb certain toxicants in order to prevent absorption in the area of gastrointestinal tract which cannot be reached by gastric lavage. The **laxatives** increase the transit of the toxic agents and prevent absorption in the lumen of the intestine. In cases of identity of the toxic chemicals, specific agents can be used to neutralize the substances.

iii. *Acceleration of elimination of toxicants*   As the consequence of absorption, the level of toxic substance attains an equilibrium in various compartments of the body such as plasma, body fluids, etc. Therefore, the withdrawal of the toxicant from one compartment will lead to the reduced concentration in another compartment. This may be accomplished by accelerating the elimination process of the chemical from the body. There are various methods to induce the rate of elimination of toxicants but appropriate method must be chosen depending on the properties of the toxicant and nature of intoxication. **Forced diuresis** increases renal clearance of the toxicants by increasing urine production. By altering the pH of the urine, the concentration of some substances dissolved in the filtrate can be raised, thus increasing their excretion. Many toxicants especially hydrophilic and low-molecular weight substances can be eliminated from the plasma by **haemodialysis** or **peritoneal dialysis** in which a concentration gradient for the toxicant is maintained between plasma on one side of the membrane and the rinsing fluid on the other side. This concentration gradient causes the substances to diffuse across the membrane into the rinsing fluid.

Haemoperfusion is another technique similar to that of dialysis but the blood is passed through a column, the end of which containing adsorbent particles. By this method, the substances including lipophilic compounds can be cleared at high rate. In **exchange transfusion method**, the blood from the intoxicated individual is drained out from one route and the fresh blood is introduced through another route. This will gradually reduce the concentration of the toxicant in the blood. **Chelators** are organic substances which form stable, complex heavy metals called **chelates**. They prevent the binding of metallic ions to tissues and are water-soluble, non-degradable, less toxic than the free metal ions and easily eliminated by the kidneys. Some toxicants or their harmful metabolites circulate through the enterohepatic cycle. These can be adsorbed in the lumen so that they are made non-available for reabsorption.

By inducing the liver enzymes, biotransformation process of certain toxicants can be increased. The substances which are absorbed and eliminated through the lungs can be removed more rapidly by **hyperventilation. Symptomatic treatment** is the treatment of symptoms which involves maintaining and close monitoring of vital functions in order to eliminate the toxic substances from the body.

Administration of Antidotes

Antidotes are chemical substances which reduce the effect of intoxication by reacting with the toxicants. Though antidotes are used to treat many intoxications, the specific available antidotes are quite lesser in number. The effect exerted by the antidotes is more or less a type of antogonism. The activity of toxic substances can be altered by chemical antagonism as in the case of chelators involving physiological and biochemical mechanisms. A few antidotes against intoxication are given in Table 24.3.

Table 24.3   Antidotes and their mode of action against intoxication

Intoxication	Antidote	Mechanism
Organophosphate and carbamate pesticides	Atropine	Antagonistic
Atropine	Physostigmine	A reversible AchE inhibitor
Opiates	Naloxone	Antagonistic
Methaemoglobulin formation	Methylthione	Reversible oxidation reduction
	Ascorbic acid	Reductant
Methanol or ethylene glycerol	Ethanol	Reduction in toxic degradation product

CHELATION

Chelation is one of the most important chemical processes by which plants and animals utilize the inorganic metals. The substances which bring out chelation are called **chelating agents or chelators.** These are organic compounds (chlorophyll, haemoglobin, cytochrome *c* catalase, peroxidase and other metalloenzymes) responsible for linking together of metal ions to form complex structures called chelates. A chelator also combines with a respective toxic ion to form a complex which will possess a lesser toxicity and can be eliminated more easily from the body. This complex is highly stable but only some exchange reactions are possible. The large complexes with positive charges will cross the membranes very slowly. The chelating agents contain more than one functional group, so they link all the coordination positions of a metallic ion. Thus the linkage between the metal ion and the chelator is by coordination similar to covalent bonds. However, the electrons for the link are supplied by the binding atoms such as H, Na, Mg, Cu, Zn as well as Mn, Fe and Co. Therefore, the resulting compound is called "metal complex" or coordinated compound. There are number of endogenous chelating agents like glutathione in the body of animals for a variety of toxic metals. They play a role in the excretion and removal of certain toxic metals. Some of the selected chelating agents are given below.

EDTA (Ethylenediamine tetraacetic acid) It is a chelating agent used in metal poisoning and is able to form complexes with several bivalent and trivalent metals. For example, calcium-disodium methylenediamine tetraacetic acid (Ca Na$_2$ EDTA) is absorbed into the gastrointestinal tract and is used for the treatment of lead poisoning with the elimination rate more than 80%. But this chelator is found to cause a variety of side effects such as hypocalcaemia, necrosis of renal tubular cells, haematuria, proteinuria, chelation of essential metals, teratogenicity, redistribution of metals to the brain, and so on.

***Dimercaprol or British Anti Lewisite* (BAL)**   It is highly liposoluble and permeates all the body tissues. It is used as an antidote to Lewisite which is excreted as dithiols and glucuronides. The side effects of BAL include increased arsenic and mercury burden in tissues, increased systolic and diastolic blood pressure, vomiting, headache, rhinorrhoea, salivation, profuse sweating, chest and abdominal pain, anxiety, etc.

D-*Penicillamine*   It is a hygroscopic crystalline powder and permeates into the cell membrane, gets metabolized and excreted very fastly through urine. However, it reduces transaminase activity and cause hypersensitivity—allergic reactions like fever, skin rashes, leukopenia, nephro-toxic effects, nausia and vomiting.

***Demercaptosuccinic acid* (DMSA)**   It is orally active and less toxic than BAL and is used in lead poisoning. It is capable of binding with albumin and distributed only in extracellular fluid with an elimination rate 30 times higher than BAL. The side effects caused by this substance include increase in transaminase activity, gastrointestinal discomfort, mild neutropenia, elevated liver enzymes, embryo and foetal toxicity, etc.

***Dimercaptopropane 1-sulphonate* (DMPS)**   It is distributed in extracellular spaces as well as inside the cells without involving in important metabolic pathways and is eliminated rapidly through kidneys. It exerts major adverse pathological effects but it is found to bring alterations in copper content in certain organs.

Deferroxamine **(DFO)** This substance is produced by *Streptomyces* distributed extracellularly as it is poorly absorbed in the gastrointestinal tract and is readily excreted in the urine. The side effects caused by this substance are hypotension, auditory and ophthalmic toxicity and bacterial and fungal infections.

Diethylenetriamine pentaacetic acid **(DTPA)** It is used to link plutonium, cobalt, zinc, etc. and can be excreted quickly from the body through kidney with 90–100% elimination rate. It is found to cause nephrotoxicity and foetal mortality.

Triethylenetetraamine **(Trien)** It is free from any side effects and is used to link divalent cations. It is excreted through urine at 50% rate and through faeces at 20% rate.

Chelation Therapy

The chelating agents are mainly used to remove the effects of toxic metals in man and experimental animals. In other words, it is the method of reducing the body burden of metals. However, most of the currently used chelators cause serious adverse side effects so that they possess only lesser therapeutic benefit. In order to minimize the problems of chelators, "adjuvants" such as essential metals, vitamins, amino acids, etc. are used in practice. The Table 24.4 explains certain specific chelating agents in the treatment of metal poisoning.

Table 24.4 Chelating agents in metal poisoning with recommended dose

Metal poisoning	Antidotes	Recommended dose	No. of days
Arsenic	BAL	2.5 mg/kg	3
	DMSA (in mild poison)	10 mg/kg/8 h	5–7
		10 mg/kg or 30 mg/kg	Further 10–14
	(in severe poison)	300 mg/kg/6 h	5
	DMPS (in mild poison)	100 mg/kg/8 h	3
Iron and aluminium	DFO	Not determined	10–12
Cadmium	DTPA supplemented with Zn	40 – 80 mg/kg or 20 – 60 mg/kg	7 7
Lead	EDTA and EDTA + BAL (in children)	75 mg/kg/day (in 2 or 3 doses)	–

DMSA	100 mg/kg/day	5
(in chronic condition)		
(In children)	30 mg/kg/day (in 3 doses)	5
(In adults)	250 mg/kg/4 h	1st day
	+	
	250 mg/kg/6 h	2nd day
Mercury BAL	Not favoured	–

REVIEW QUESTIONS

1. What is intoxication? Explain the different types of intoxication processes.

2. Explain the following:
    i.   Accidental intoxiation
    ii. Deliberate intoxication
    iii. Chronic intoxication

3. Give an account of occupational poisoning.

4. Describe the preventive measures to reduce absorption of toxicants.

5. What is gastric lavage?

6. In what ways, the elimination of toxicants can be accelerated?

7. Write an account of antidotes.

8. Explain the various antidotal procedures to terminate the effect of toxic substances.

9. What is meant by chelation? Explain its mechanism.

10. Write an essay on chelating agents and their uses.

11. Why does the chelating agents possess very low therapeutic use?

25

SELECTED
TOXICOLOGICAL METHODS

EVALUATION OF TOXICITY OF A POLLUTANT THROUGH LC$_{50}$ 96 HR VALUE IN AQUATIC ORGANISMS

Aim

To determine LC$_{50}$ 96 hr value of a pollutant by using aquatic organisms as test animals.

Materials

Selected pollutant (industrial effluents or pesticides or heavy metals, etc.), live test animals (freshwater animals such as dragonfly larvae, fish or crab), plastic troughs and diluent medium.

Method

Determination of LC$_{50}$ 96 hr value of a pollutant to an organism is a typical test used to evaluate the toxicity of a pollutant in which groups of test animals are reared in different concentrations of the pollutant for a specific period of time and the number of dead and alive animals are scored.

1. *Selection and maintenance of test animals in the laboratory*   The test animals are collected from their breeding places, brought to the laboratory and acclimatized in tap water contained in plastic troughs for a week. During this period, the animals are fed with their favourite food and the medium is replaced daily.

2. *Selection of pollutant and diluent medium*   The pollutant which is to be tested is chosen. If the pollutant is a liquid, the undiluted one is considered to be 100%. Various concentrations are prepared by diluting it using dechlorinated tap water (diluent medium). If the pollutant is a solid, a stock solution of 100 mg/ml is prepared in an organic solvent and is used to prepare various concentrations by using dechlorinated tap water. In the test water, variables such as temperature, pH, salinity and dissolved oxygen are controlled.

3. *Determination of* LC_{50} *96 hr value*   A group of 10 laboratory acclimatized test animals having the same weight, size and age are introduced into each test concentration of the pollutant contained in plastic troughs (four or five concentrations ranging from low to high are to be selected). Appropriate controls are also maintained simultaneously for each pollutant concentration. The test animals are not fed during the experimental period. The test solutions are replaced daily by appropriate concentrations in each container. The number of dead and alive larvae are scored after a specified time interval (96 hr) in each concentration. A minimum of three or four replicates are maintained in each concentration and the average mortality in each concentration is taken into consideration. The percentage kill of the test animals is converted into probit kill by using the conversion table given by Finney (1964). The data are tabulated and graphically represented.

The LC$_{50}$ value thus obtained by graphical method can be verified by the following formula:

$$\log LC_{50} = \frac{\log A50 - A}{B - A} \times \log 2$$

where,

A = Concentration of the pollutant which kills 50% of the test animals,

a = % of mortality immediately below 50%

b = % of mortality immediately above 50%

Table 25.1 Transformation of percentages to probits

%	0	1	2	3	4	5	6	7	8	9
0		2.67	2.95	3.12	3.25	3.36	3.45	3.52	3.59	3.66
10	3.72	3.77	3.82	3.87	3.92	3.96	4.01	4.05	4.98	4.12
20	4.16	4.19	4.23	4.26	4.29	4.33	4.36	4.39	4.42	4.45
30	4.48	4.50	4.53	4.56	4.59	4.61	4.64	4.67	4.69	4.75
40	4.75	4.77	4.80	4.82	4.85	4.87	4.90	4.92	4.95	4.97
50	5.00	5.03	5.05	5.08	5.10	5.13	5.15	5.18	5.20	5.23
60	5.52	5.55	5.58	5.61	5.64	5.67	5.71	5.74	5.77	5.81
70	5.52	5.55	5.58	5.61	5.64	5.67	5.71	5.74	5.77	5.81
80	5.84	5.88	5.92	5.95	5.99	6.04	6.08	6.13	6.18	6.23
90	6.28	6.34	6.41	6.48	6.55	6.64	6.74	6.88	7.05	7.33
99	7.33	7.67	7.41	7.46	7.51	7.58	7.65	7.75	7.88	8.09

EVALUATION OF TOXICITY OF
A POLLUTANT THROUGH LC$_{50}$ 96 HR
VALUE IN TERRESTRIAL ORGANISMS

Aim

To determine LC$_{50}$ 96 hr value of a pollutant by using terrestrial organisms as test animals.

Materials

Selected pollutant, test animals (terrestrial insects such as cockroach or *Gryllus*), glass jars and diluent medium.

Method

While using terrestrial animals as test organisms, they are exposed to the pollutants by contact method or topical application (either dipping or spraying) for a specific period of time.

The experimental animals and various concentrations of the pollutant are selected as in the toxicity test for aquatic organisms.

In **spraying method** different concentrations of each pollutant is sprayed daily on the surface of the body (excluding the wings) of a group of 10 laboratory acclimatized animals with the help of a syringe and are kept in separate cages.

In **dipping method**, the filter paper is dipped in each concentration of the pollutant and is kept in glass jars as an inner layer and a group of 10 laboratory acclimatized animals are introduced into the container. The filter paper is made wet by appropriate concentration of the pollutant daily. After a specified time interval (96 hr), the mortality of the test animals in each concentration with adequate replicates are scored, converted into percentage and probit kill was tabulated and graphically represented.

Table 25.2 Mortality of test animals in different concentrations of a pollutant after 96 hr exposure period

Concentration of the pollutant (%)	No. of animals exposed	No. of animals dead	% kill	Probit kill (%)	LC_{50} 96 hr
0	10	0	0	0	
5	10	0	0	0	
10	10	3	30	4.40	
15	10	5	50	5.00	15%
20	10	6	60	5.25	
25	10	9	90	9.28	
30	10	10	100	8.09	

DETERMINATION OF THE EFFECT OF TEMPERATURE ON THE TOXICITY OF A POLLUTANT

Aim

To determine the effect of temperature on the toxicity of a pollutant.

Materials

Rectangular jars, beaker, thermometer, selected pollutant, test animals (fishes or dragonfly larvae), hot water, cold water, etc.

Method

1. A healthy test animal of known weight is exposed to the pollutant contained in a beaker. The time at which the fish dies is recorded (survival time) at room temperature.

2. Another test animal of the same weight is exposed to the pollutant in a beaker which is kept in a water bath.

3. The temperature of the pollutant medium is increased by 2°C above the room temperature by adding hot water to the water bath. Now the survival time of the animal is recorded.

4. Then a beaker containing the pollutant along with a third animal is kept in a water bath.

5. The temperature of the pollutant medium is decreased by 2°C below the room temperature by adding cold water and the survival time of the animal is recorded. The results are tabulated and graphically expressed.

Table 25.3   Effect of temperature on the toxicity of a pollutant

Temperature (°C)	Survival time of the test animal (min.)
Room temperature	a
2°C above room temperature	b
2°C below room temperature	c

DETERMINATION OF THE EFFECT OF pH ON THE TOXICITY OF A POLLUTANT

Aim

To determine the effect of pH on the toxicity of a pollutant.

Materials

Rectangular jars, beakers, selected pollutants, test animals, pH tablets of pH 7.0, 9.2, etc.

Method

1. A healthy test animal of known weight is exposed to a pollutant of known pH (8.0) contained in a beaker.

2. The survival time of the test animal is recorded at this pH level of the pollutant.

3. The pH of the pollutant is decreased by using pH tablet 7.0 and another test animal is introduced into it.

4. Then the pH of the pollutant medium is increased by using pH tablet 9.2 and a third animal is kept in it.

5. In both the cases, the survival time of the animal is noted, tabulated and represented graphically.

Table 25.4   Effect of pH on the toxicity of a pollutant

pH of a pollutant	Survival time of the test animal (min)
7.0	a
8.0	b
9.2	c

DETERMINATION OF COMBINED TOXICITY OF POLLUTANTS ON AQUATIC ORGANISMS

Aim

To evaluate the nature of combined toxicity of two different pollutants on aquatic organisms.

Materials

Selected pollutants, freshwater test animals, plastic troughs and diluent medium.

Method

1. Static bioassays are carried out to determine LC_{50} 96 hr values for two different pollutants by using the laboratory acclimatized animals with appropriate replicates.

2. The concentration of the individual pollutant which killed 50% of the test animals is prepared separately (stock solution).

3. By using the stock solutions, different ratio of combinations of the pollutants are prepared.

4. A group of 10 laboratory acclimatized test organisms are introduced into each test combination for a 96 hr duration.

5. The average mortality of the animals at each ratio is converted into percentage kill and then into probit kill.

6. After the determination of LC_{50} 96 hr values for the two toxicants at various ratio of combinations, parameters like upper and lower confidence limits, safe level, co-toxicity coefficient, index "V" value and linear "S" index are calculated to determine the nature of interaction.

QUALITATIVE SEPARATION OF ORGANOCHLORIDES AND ORGANOPHOSPHATES

Aim

To separate and analyse the residues of organochlorides and organophosphates by thin layer chromatography (TLC).

Materials

Samples to be analysed, TLC plates, drier, desiccator, separating funnel, volumetric flasks, centrifuge, elution tubes, etc.

Method

1. *Sampling and storage of samples* Samples like water, fatty and non-fatty foods, vegetables, fruits, etc. are selected at random and stored in a freezer below 0°C.

2. *Extraction and preparation of residue* 200 ml of acetonitrile is added to an approximate quantity of the sample. Then it is blended well and centrifuged at 2000 rpm for 5 minutes. The residue is again used to get the extract and the two acetonitrile extracts are combined. In a separating funnel (1 litre), 100 ml of acetonitrile extract is taken and 200 ml of distilled water, 10 ml of dichloromethane, 50 ml of saturated sodium sulphate, and 200 ml of *n*-hexane are added and shaken well. Then the hexane layer is withdrawn in a Petri dish. This is repeated for four times with 50 ml *n*-hexane. The hexane extracts are pooled, 20 g of anhydrous sodium sulphate is added to remove water and concentrated to 5 ml in an evaporator.

3. *Elution* A glass tube (30 cm long and 2.5 cm diameter) is packed with florisil in hexane to about 15 cm height.

Anhydrous sodium sulphate is placed over the florisil column to about 2.5 cm height. The concentrated hexane extract is slowly added over it and eluted with solvent systems to separate the pesticides into various groups.

Pesticides	Eluting solvent
Aldrin, DDT, heptachlor	100 ml *n*-hexane
Endosulphan, lindane, endrin, dieldrin, methoxychlor	200 ml 6% diethyl ether in *n*-hexane
Malathion, parathion, diazinon	200 *n*-hexane ml 15% diethyl ether in *n*-hexane

4. *Detection of organochlorides*   Glass plates of 20 × 10.5 cm are coated with 450 µm layer of silica gel G (in water) slurry, air-dried, activated in 110°C and stored in a desiccator. The concentrated sample extract is applied on TLC plates, developed in a solvent system and air-dried. The plates are sprayed with the chromogenic reagent (1% orthotoluidine in acetone), air-dried and exposed to sunlight.

DDT and its breakdown products appear greenish blue in colour.

5. *Detection of organophosphates*   The TLC plates are spotted with the concentrated sample extract and developed in acetone–hexane(1 : 4). The plates are sprayed with the chromogenic reagent (0.5% brilliant green), air-dried and exposed to bromine vapour for 30 seconds.

The organophosphates appear as yellow spots.

QUALITATIVE SEPARATION
OF CARBAMATE PESTICIDES

Aim

To separate and analyse the residues of carbamate pesticides by thin layer chromatography.

Materials

Samples to be analysed, TLC plates, drier, desiccator, separating funnel, volumetric flasks, centrifuge, elution tubes, etc.

Method

1. *Extraction*   The residue is extracted from the sample with methylenechloride in 2 : 1 solvent–substance ratio. If the residue contains impurities, acetone acetonitrile is used as the solvent. The animal tissues are homogenized for the extraction of the residue. In water, the residue is extracted with methylchloride in a separating funnel.

2. *Cleaning*   The extract obtained is treated with animal charcoal or C190 N Nuchar and filtered. The filtrate is cleaned by any one of the following three methods:

  i. *Coagulation*   To the filtrate, an aqueous solution of 0.1% ammonium chloride and 0.2% phosphoric acid are added and filtered. The supernatant is used.

  ii. *Column chromatography*   By using florisil as the column adsorbent and methylenechloride as the eluting solvent, the residue is extracted.

  iii. *Partitioning*   By using an immiscible polar and nonpolar solvent systems, the residue can be partitioned into the polar solvent.

3. *Detection*   A slurry of 25 g silica gel in 50 ml 0.5% H_2SO_4 is coated (250 μm thickness) on 20 × 20 cm glass plates, air-dried and activated at 110°C for 30 minutes. After spotting the sample, the plates are developed with one of the solvent systems such as benzene and acetone (95:5 or 85:15), cyclohexane and ethanol (85:15), hexane, toluidine and acetone (60:20:20), and hexane and acetone (70:30), air-dried, heated in an oven for 15 minutes, sprayed with dimethylaminobenzaldehyde and reheated.

The residue is confirmed by the R_f value in the appropriate solvent system.

ANALYSIS OF PESTICIDE RESIDUES BY FINGER PRINTING TECHNIQUE

Aim

To evaluate qualitatively the presence of pesticide residues in vegetable samples.

Materials

Whatman No.1 filter paper, vegetable samples, *o*-toluidine, acetone, etc.

Method

1. The vegetable samples procured from local markets of different areas are made into transverse or longitudinal sections of 2 cm thickness.

2. Whatman No.1 filter paper which is sprayed with 1% *o*-toluidine in acetone, dried and stored in the dark serves as a chromogenic paper to detect the residues.

3. The individual section of vegetables are pressed against the chromogenic paper for 30 seconds to allow *in situ* reaction of the cell-bound pesticide residues.

4. Then the paper is exposed to sunlight so as to form pesticide fingerprints instantly on the paper.

5. The detection and location of various pesticides are made by referring to the standard colours developed by the individual pesticides on the chromogenic paper.

Pesticides	Colour developed
Organochlorines	
DDT	Light green or leafy green
HCH	Prussian violet
Aldrin	Yellowish green
BHC	Light orange
Endosulphan	Yellow
Organophosphates	
Monocrotophos	Dark brown
Carbamates	
Carbaryl	Brown
Carbofuron	Violet

EVALUATION OF SUBLETHAL TOXICITY OF A POLLUTANT USING AQUATIC ORGANISMS AS EXPERIMENTAL ANIMALS

Aim

To evaluate the chronic effect of a pollutant in aquatic animals.

Materials

Selected pollutant, test animals, plastic troughs, diluent medium, etc.

Method

1. After the determination of the LC_{50} 96 hr value of a pollutant to the test animals, a few (atleast 5) different sublethal concentrations of the pollutant are prepared.

2. Generally, 1/10th of the concentration of the LC_{50} 96 hr value is considered to be the ideal concentration for the chronic exposure.

3. Six experimental containers are set up. Of them, five contain different concentrations of the pollutant with ten animals in each. The remaining one containing tap water with 10 test animals is kept as control.

4. The test animals are fed daily throughout the experimental period and the test solutions are replaced daily by appropriate concentrations in each container.

5. After a stipulated period of time (a minimum of 3 weeks), the test animals are used for histopathological, physiological, biochemical and enzyme studies.

OBSERVATION OF HISTOPATHOLOGICAL ALTERATIONS IN POLLUTANT-TREATED ANIMALS

Aim

To observe histopathological alterations in tissues of animals treated in a pollutant.

Materials

Beakers, scissors, cover slips, forceps, coplin jars, microslides, microtome, paraffin bath, specimen tubes, slide warmer, alcohol, acetone, eosin, formalin, glacial acetic acid, haemotoxylin, HCl, Mayer's albumin, methyl benzoate, paraffin, sodium salicylate, xylene, Ringer's solution, *n*-butyl alcohol, phenol, etc.

Method

1. From each test animal, in each concentration of the pollutant, the tissue is dissected out in Ringer's solution for processing.

2. The tissue is fixed in the Khale's fluid (95% ethanol : formalin : acetic acid (15 : 6 : 1 v/v/v)) for three h.

3. Then it is processed by keeping the following solution mixtures stepwise:

   Water : ethanol : *n*-butyl alcohol (3 : 5 : 2 v/v/v) — 1 h.

   Water : ethanol : *n*-butyl alcohol (1 : 8 : 11 v/v/v) — 1 h.

   Water : ethanol : *n*-butyl alcohol : phenol (11 : 50 : 35 : 4 v/v/v/v) — 24 h.

   Ethanol : n-butyl alcohol (1 : 3 v/v) — 1 h.

4. The tissue is kept in 4% phenol in *n*-butane for 2 hours and subjected to cold infiltration for about 16 h by using equal volume of molten paraffin wax (melting point (58–60°C) and 4% phenol in *n*-butanol.

5. It is followed by hot infiltration in which the material is transferred through three changes to fresh molten wax for a period of 3 h (1 h in each change).

6. After embedding in paraffin wax, sections of 8 micron thickness are cut and are stained with Delafield's haematoxylin using eosin as counter stain.

7. The stained sections are microphotographed and observed under the microscope for histopathological alterations in comparison with the control tissues.

EVALUATION OF THE IMPACT OF A POLLUTANT ON THE HAEMOPOIETIC ACTIVITY

Aim

To evaluate the haemopoietic activity in a fish that is exposed to a pollutant by observing the activity of bone marrow and spleen.

Materials

Control and pollutant-treated fish dissection instruments, mortar and pestle, slides, microscope, physiological saline, methanol, Leishman's stain, etc.

Method

1. The fish is stunned by a blow on the head.

2. A fragment of the vertebral column (approximately 2 cm) and spleen are dissected out, crushed in a mortar and pestle with few drops of physiological saline and the extracts are used.

3. A drop of the extract is placed near the end of a clean and dry slide and the smear is made exactly in the similar way as the blood smear.

4. The smear is air-dried and absolute methanol is added and incubate it for 20 minutes.

5. The smear is stained with Leishman's stain and examined under the microscope for myeloid, erythroid and reticulocytes.

6. An average percentage value is calculated for each cell type.

7. Then the smear is microphotographed.

DETERMINATION OF OXYGEN CONSUMPTION IN THE DRAGONFLY LARVAE EXPOSED TO A POLLUTANT

Aim

To determine the rate of oxygen consumption in small aquatic insects like dragonfly larvae treated with a pollutant by using specially designed respiratory chamber.

Materials

Control and pollutant-treated dragonfly larvae, respiratory chamber, reagent bottles, burette, pipette, conical flask, manganous chloride, alkaline iodide, concentrated H_2SO_4, 1% starch solution, sodium thiosulphate solution (0.015N), etc.

Method

1. *Designing of respiratory chamber* The mouth of a wide-mouthed 500 ml bottle is fitted with a four-holed rubber stopper. One hole is to insert a centigrade thermometer and the remaining three for glass tubes with pinchcocks serving as delivery tubes. One tube is to draw water into the respiratory chamber, the second is to test if the chamber is air-tight and the third is to drain water from the chamber. The chamber is coated with black paint to prevent the entry of light into the chamber.

2. *Determination of oxygen consumption* A healthy dragonfly larva of known weight is placed in the respiratory chamber containing water. Immediately, the initial sample is collected from the chamber and the time is noted. The animal is left for 1 h and then the final sample is collected from the chamber. The oxygen content in the initial and final water samples is determined by Winkler's method. The initial oxygen content minus the final oxygen content of the water accounts for oxygen consumed by the animal in one hour. The rate of oxygen consumed by the animal is calculated by using the following formula:

$$\text{Rate of oxygen consumption} = \frac{\text{Oxygen consumed by the animal in one hour}}{\text{Weight of the animal (g)}} \text{ ml of } O_2/g/h$$

ESTIMATION OF ORGANIC CONSTITUENTS IN THE HAEMOLYMPH OF DRAGONFLY LARVA

Aim

To estimate various organic constituents present in the haemolymph of a dragonfly larva.

Method

Haemolymph is aspirated into the watch glass by means of a 18-gauge hypodermal needle and syringe through a puncture made in the mid-dorsal line of the trunk segment. Coagulation is prevented by adding citrate salts. Since the haemolymph quantity from a single specimen is insufficient for analyses, pooling of haemolymph from few specimens of the same state are made. Before collecting the haemolymph, the larvae are starved for 24 h to eliminate the effect of differential feeding on the biochemical constituents of the haemolymph.

Sample preparation The haemolymph of known volume (0.05 ml) is drawn and is centrifuged for 20 minutes at 2500 rpm to separate the haemocytes. The supernatant is used for the biochemical estimations.

1. *Estimation of total free amino acids* To 0.05 ml of haemolymph 10 ml of 10% trichloro acetic acid is added to precipitate the proteins and centrifuged for 30 minutes at 5000 rpm to separate the precipitated proteins. To one volume of the supernatant, two volumes of ninhydrin reagent is added and kept in a boiling water bath for 6.5 minutes along with standard without haemolymph sample. It is cooled under running tap water. After cooling, the OD is measured at 570 nm in a spectrophotometer. The amount of free amino acids is determined by the following formula:

$$\frac{OD\,of\,sample}{OD\,of\,standard} \times \frac{0.006}{0.025} \times 100$$

where,

0.006 — Amount of nitrogen present in the standard

0.025 — Amount of extract present in 1ml of the sample.

2. *Identification of differential free amino acids* Two-dimensional paper chromatographic method is adopted to identify various amino acids in the haemolymph.

Preparation of sample To 0.05 ml of haemolymph in a test tube, three volumes of 80% ethanol is added and centrifuged for 30 minutes at 5000 rpm to precipitate the proteins. To the supernatant, three volumes of chloroform are added. The resulting aqueous layer formed at the top is used for spotting.

Identification   Whatman No.1 paper is cut into squares of 12 cms × 12 cms. The solvent for the first run is *n*-butanol and 3% ammonia (150 : 60 v/v) and for the second run is *n*-butanol, acetic acid and water (12 : 3 : 5 v/v/v). After the run, the chromatograms are dried at room temperature. The amino acids are visualised by spraying 0.1% ninhydrin in *n*-butanol (w/v) and dried at 100°C for 3 minutes. The amino acids are identified by comparing with standard chromatogram. The developed chromatograms are photographed.

3. *Estimation of total proteins*   To 0.05 ml of the haemolymph, 5 ml of 80% ethanol is added and centrifuged for 15 minutes at 5000 rpm. To 2 ml of the supernatant, 3 ml of Biuret reagent is added and kept in a boiling water bath at 37°C for 10 minutes. It is cooled, and the OD is measured at 540 nm in a spectrophotometer. The amount of protein present in haemolymph is obtained by referring to the standard protein curve.

4. *Electrophoretic separation of haemolymph proteins*

Preparation of sample   To 25 µl of the haemolymph, 0.5 ml of 2% sodium dodecyl sulphate (SDS) solution containing each 0.25 ml of 5% mercaptoethanol and 10% sucrose solution are added prior to electrophoreris. This mixture is kept in a boiling water bath for 90 seconds and then in ice until loading. The SDS denatures protein by binding with individual polypeptides and mercaptoethanol disrupt the disulphide bonds to keep the peptides apart.

Separation   Using a 10 ml pipette, the separating gel mixture is poured immediately into the glass sandwich up to the mark with no air bubbles. Immediately after pouring the separating gel mixture, 0.5–1 ml of the overlay solution is injected slowly. The overlay solution spreads over the entire top surface of the separating gel mixture. This is kept undisturbed for polymerization (15–20 minutes). Then, the glass sandwich is tilted and the overlaid solution is discarded. The comb is inserted and the stacking gel mixture is poured without air bubbles. Polymerization is allowed to take place for about 15–30 minutes. The comb is removed carefully and the wells are washed with tank buffer. Using a Hamilton syringe, 25 μl of the sample is applied to each well. The top chamber is filled with tank buffer and the wells are also filled with buffer very gently using a syringe. An equal volume of the tank buffer is added to the bottom reservoir. The tank is connected to the power supply and is set to constant current mode (10–15 mA current). The run is continued for 90 minutes till the marker dye reaches the bottom of the gel. The upper reservoir is drained and the gel is taken out carefully and stained for 3–4 h. After staining, it is destained several times till a clear background is obtained. Then the bands are examined in a white light transilluminator and photographed. The densitometric tracing of haemolymph proteins is done by gel documentation system.

5. *Estimation of total free sugars* To 0.05 ml of haemolymph, 5 ml of 80% ethanol is added and centrifuged at 3000 rpm for 5 minutes. To 1 ml of the supernatant, standard and blank, 10 ml of anthrone reagent is added and heated in a water bath for 10–15 minutes. The OD of the sample and standard are read at 620 nm. The concentration of total free sugars is calculated by using the following formula.

$$\frac{\text{OD of the unknown}}{\text{OD of standard}} \times \frac{\text{Concentration of}}{\text{standard}} \times \text{Dilution factor}$$

6. *Estimation of reducing sugars* To 0.05 ml of haemolymph, 1.8 ml of 80% ethanol is added and centrifuged for 5 minutes at 3000 rpm. The supernatant is made up to 5 ml by adding distilled water. To 0.05 ml of the sample, 0.5 ml of standard glucose is added along with 1 ml of alkaline copper reagent. The top of the bottle is covered with glass marble. The mixture is heated in a water bath for 20 minutes and cooled at room temperature. Then 1 ml of arsenomolybdate colour reagent is added followed by 5 ml of distilled water. Optical density is read at 620 nm.

7. *Estimation of non-reducing sugars* The quantity of non-reducing sugar is calculated by subtracting the quantity of reducing sugars from total sugars.

8. *Estimation of glycogen content* To 0.05 ml of the haemolymph, 5 ml of 80% methanol is added and are centrifuged at 3000 rpm for 5 minutes. To the residue, 5 ml of 5% TCA is added and boiled in a water bath for 5 minutes. The quantity is made to 5 ml by the addition of 5% TCA and centrifuged. To 2 ml of the supernatant, 6 ml of concentrated sulphuric acid is added and heated in a boiling water bath for 6.5 minutes and then cooled. The OD is read at 530 nm. A standard graph is made by using glucose as the standard.

9. *Estimation of total lipids* To 0.05 ml of haemolymph, 5 ml of chloroform–methanol mixture is added and allowed to stand for sometime and a few ml of 0.05N KCl is added. It is allowed to stand for few minutes and the aqueous phase is discarded. The remaining chloroform phase is transferred to a weighed container and is dried in a vacuum desicator for one or two days. The difference in the weight of the container represents the lipid content of the haemolymph.

10. *Estimation of cholesterol* The dried lipid extract residue is dissolved in 2 ml of acetyl chloride and mixed well. The test tube is covered with a clean glass marble and kept in a water bath at 65°C for 15 minutes. Then the tube is cooled down to room temperature. The intensity of red colour is measured at 528 nm in a spectrophotometer. The concentration of cholesterol is estimated by using a standard graph.

11. *Estimation of triacyl glycerol (TAG)* Haemolymph sample (0.05 ml) is centrifuged and the supernatant is taken. To 0.05 ml of the supernatant, 4 ml of isopropanol and 400 mg of washed alumina are added and mixed thoroughly in a mechanical rotator for 15 minutes. The mixture is centrifuged and the supernatant is collected and added with 0.6 ml of saponifying agent. The mixture is incubated at 60°C for 15 minutes and cooled. To this, 1 ml of metaperiodate and 0.5 ml of acetyl acetone are added and incubated at 50°C for 30 minutes. The OD of the sample is read out at 405 nm. The TAG is calculated by using the following formula:

$$\frac{\text{OD of the unknown}}{\text{OD of the known}} \times 300$$

where, 300 refers to the concentration of standard.

All the biochemical parameters are expressed in mg/100 ml haemolymph.

ESTIMATION OF INORGANIC CONSTITUENTS IN THE HAEMOLYMPH OF DRAGONFLY LARVA

Aim

To estimate various inorganic constituents present in the haemolymph of a dragonfly larva.

Method

Preparation of sample Haemolymph of known volume (0.05 ml) is withdrawn from the insect and centrifuged for 5 minutes at 3000 rpm. The supernatant is decanted for analysis.

1. *Estimation of sodium* The sodium content in the haemolymph is estimated by using flame photometer. The supernatant is to with 10 ml of distilled water and OD is recorded. The concentration of sodium in the sample is determined from a standard graph by using 100 ppm standard NaCl solution. Then the concentration is converted into mEq/litre as given below,

Concentration of Na in the sample = A ppm

In terms of mEq/l = $A/23$

where, 23 is a constant.

2. *Estimation of potassium* The concentration of potassium in the haemolymph is also determined with the help of flame photometer as in Na content and the concentration is also represented as mEq/litre.

Concentration of K in the sample = A ppm

In terms of mEq/l = $A/39$

where, 39 is a constant.

3. *Estimation of calcium and magnesium*

Calcium 5 ml of the sample is taken in a porcelain basin and 10 ml of 10% sodium hydroxide solution and 50 mg of murexide indicator are added. This mixture is titrated against 0.002N EDTA solution till the colour changes from pink-red to purple or violet. The volume of EDTA consumed is recorded.

1 ml of 0.002N EDTA contains 0.00004 g of Ca.

A ml of 0.002N EDTA contains 0.00004 × A g of Ca.

$$\text{In terms of mEq}/1 = \frac{0.00004 \times A}{\text{Volume of sample}} \times 1000 \times \frac{1000}{20.0}$$

Magnesium To 5 ml of the sample, 10ml of ammonium chloride–ammonium hydroxide buffer solution and 2–3 drops of eriochrome black-T indicator are added. This is titrated against 0.002N EDTA solution till the colour changes from wine red to sky blue. The volume of EDTA consumed is noted. The calcium and magnesium content were calculated in the following manner:

Volume of sample taken = 5 ml

Volume of EDTA consumed for Ca + Mg = A ml

Volume of EDTA consumed for Ca alone = B ml

Volume of EDTA consumed for Mg alone = $(A–B)$ ml

1 ml of 0.002N EDTA contains 0.000024 g of Mg.

$(A–B)$ ml of 0.002N EDTA contains 0.000024 × $(A–B)$ g of Mg.

$$\text{In terms of mEq}/1 = \frac{0.000024 \times (A-B)}{\text{Volume of sample}} \times 1000 \times \frac{1000}{12.16}$$

4. *Estimation of chlorides*   5 ml of the sample is pipetted out into a clean conical flask and few drops of potassium chromate indicator are added. It is titrated against 0.002 N silver nitrate solution till the appearance of flesh red coloured precipitate. From the volume of silver nitrate solution consumed, the chloride content is calculated as follows:

Volume of sample = 5 ml

Volume of silver nitrate consumed = *A* ml.

1 ml of 0.002N silver nitrate contains 0.000178 g

A ml of 0.002N silver nitrate contains 0.000178 × *A* g of chloride

$$\text{In terms of mEq} / 1 = \frac{0.000178 \times A}{5\,\text{ml}} \times 1000 \times \frac{1000}{35.5}$$

ENZYME ASSAYS IN THE TISSUES OF ANIMALS

Aim

To determine the activity levels of various enzymes in the tissues of dragonfly larva.

Sample Preparations

Fresh tissues (cuticle, fat body and nervous tissue), each weighing 100 mg, are separately homogenized in a few ml of water under cold conditions. Haemolymph of known volume (0.05 ml) is used. For each tissue, a 10% homogenate is prepared. Homogenates are centrifuged individually at 5000 rpm for 10 minutes. The supernatants are used as enzyme source for various enzyme assays.

1. Protease Activity

Substrate preparation An exact amount of casein is weighed and made into a suspension in water (2.5% w/v). It is dissolved by adding 0.1N NaOH drop by drop. The desired pH is maintained by adding acetic acid.

Method To 1 ml of the substrate, 1 ml of phosphate buffer and 0.2 ml of enzyme source are added. The mixture is incubated at 37°C for one hour. To this mixture, 1 ml of 12% TCA is added, cooled in ice immeditely and centrifuged at 5000 rpm for 30 minutes. The amount of protein in the precipitate is determined. A control is prepared by adding 1 ml of 12% TCA followed by 0.2 ml of the enzyme source to 1 ml of substrate. The protein content of the control is determined in the precipitate without incubating for an hour. The difference gives the amount of protein that has been utilized.

The result is expressed as μ mole casein/mg protein/hour.

Determination of optimum pH To find the optimum pH for protease activity, a number of controls, each with 1 ml of casein, 1 ml of 12% TCA and 0.2 ml of enzyme source, are made. The activity is measured under different pH by using veronal acetate buffer (pH 3.0 for acid, pH 9.0 for alkaline) and phosphate buffer (pH 7.0 for neutral). The optimum pH is found from a standard graph drawn with pH versus activity.

2. SDH Activity

Preparation of reaction mixture The reaction mixture consists of 0.1 ml of sodium succinate (0.15 M, pH 7.0), 1 ml of potassium phosphate buffer (0.4 M, pH 7.2) and 1 ml of PNP (2-*p*-iodophenyl 3-*p*-nitrophenyl 5-phenyltetrazolium chloride solution).

Method   The reaction mixture is added to 0.5 ml of enzyme source containing 0.4 ml of distilled water. The mixture is incubated at 37°C for 30 minutes and immediately thereafter, the enzyme activity is arrested by the addition of 6 ml of acetic acid. The mixture is kept in a freezer for 12 h and then the formazon formed was extracted in 6 ml of toluene. The OD is read at 495 nm. A reagent blank is prepared as above with the substitution of enzyme source with distilled water. Formazon formed in the reaction mixture is calculated by using a standard formazon graph.

The enzyme action was expressed in terms of μ moles formazon/mg protein/h.

To find the optimum pH for SDH activity, the enzyme source is assayed in different phosphate buffers ranging in pH from 6 to 9 at an interval of 0.5.

3. LDH Activity

Method   To the test tubes labelled as "test" and "blank", 1 ml of buffered substrate at pH 10 (2.7 g of lithium lactate dissolved in 0.1 ml glycine buffer) and 0.1 ml of tissue extract are added. To the test, 0.2 ml of 0.5% NAD (w/v) (nicotinamide adenine dinucleotide) is added and to the blank, 0.2 ml of distilled water. The tubes are incubated exactly for 15 minutes at 37°C. The reaction is arrested by adding 1 ml of colour reagent (20 mg of 2, 4-dinitrophenyl hydrazine dissolved in 100 ml of 1N HCl) to each tube and the incubation is continued for 15 more minutes. Then 10 ml of 0.4N NaOH is added to each tube to make the solution strongly alkaline for maximal development of colour of hydrazine formed. The colour developed is read at 440 nm. The enzyme activity is calculated from the standard curve obtained by using pyruvate standard (1 mM).

The enzyme activity was expressed as μ mole of formazon/mg protein/hour.

4. Lipase Activity

Substrate preparation   2 ml of olive oil is emulsified by the addition of 25 ml distilled water and 100 mg of bile salt (sodium taurocholate) and is used as substrate.

Method   To 1 ml of substrate, 1 ml of veronal acetate buffer of known pH and 0.2 ml of enzyme source are added and mixed well. The pH is recorded immediately. At frequent intervals (10 minutes) or as the pH drops by about 0.2 unit, 0.1N NaOH is added to bring pH to the initial value. The titration is continued for 30 minutes. The amount of NaOH added is noted and is equal to the fatty acids produced.

The values are expressed as lipase unit/mg wet tissue/hour except haemolymph which is expressed as lipase unit/ml/hour.

To determine the optimum pH, the enzyme activity is assayed at various levels of veronal acetate buffer ranging in pH from 2 to 10 at an interval of 0.5.

Acid and Alkaline Phosphatase Activity

Reaction mixture   The reaction mixture is prepared by mixing 1 ml of PNP (*p*-nitrophenol phosphate) solution with 1 ml of veronal buffer (pH 5 for acid phosphatase and pH 8 for alkaline phosphatase).

Method   The reaction mixture is mixed with 0.2 ml of enzyme source and incubated at 27°C for 1 hour. Enzyme activity is arrested by the addition of 5.9 ml of 0.05N NaOH.

The *p*-nitrophenol released is measured at 420 nm and referred to a standard graph made by using *p*-nitrophenol phosphate as the standard.

The results are expressed as mg *p*-nitrophenol formed/ mg protein/hour.

To determine the optimum pH, veronal buffer ranging from pH 2 to 10 is used at interval of 0.5.

Acetylcholine (ACh) Content and Acetylcholine Esterase (AChE) Activity

Into a test tube marked as A, 0.2 ml of the tissue extract and 2 ml of 0.5% hydroxylamine hydrochloride are taken. In another test tube labelled as B, 0.2 ml of tissue homogenate is mixed with 1 ml of 0.5% buffer acetylcholine chloride and the mixture is incubated at 37°C for 30 minutes. To this incubation mixture, 2 ml of 0.5% alkaline hydroxylamine hydrochloride is added. The contents of A and B are vigorously shaken separately and then 1 ml of 1 : 1 dilute HCl and 1 ml of 0.07 M ferric chloride are added. The brown colour developed in each sample is read at 540 nm. Another mixture without tissue homogenate serves as blank. The amount of unreacted ACh is determined from the acetylcholine chloride standard graph. The activity of AChE is determined by subtracting the amount of reacted ACh from the amount of the unreacted ACh.

The concentration of ACh is expressed as mg/100 mg wet weight of cuticle, fat body and nervous tissue and mg/100 ml of haemolymph.

The AChE activity was expressed as mg ACh hydrolysed/100 mg protein/hour.

Albumin and Globulins

To each of the test tubes marked 'S' (standard) and 'T' (test), 6 ml of sodium sulphate–sulphite solution is taken. To the tube 'T', 0.4 ml of serum sample and to the tube 'S', 0.4 ml of standard serum is added and mixed well. Immediately 2 ml of mixture from tubes S and T are transferred into the test tubes marked 'STP' (Standard total proteins) and 'TTP' (Test total proteins) respectively. The rest of the standard and test mixtures are added to 3 ml of ether and centrifuged at 3000 rpm for 5 minutes. From each of these tubes, 2 ml of supernatant is pipetted out into test tubes marked 'SA' (Standard albumin) and 'TA' (Test albumin) respectively. The tubes STP, SA, TTP and TA are kept at room temperature for 15 minutes. The OD of the above mixtures are measured separately in a colorimeter at 550 nm against blank (B). From the OD values, the quantity of total proteins, albumin, and globulin in the serum are calculated.

$$\text{Serum total proteins } (X) = \frac{\text{OD of test}}{\text{OD of standard}} \times \text{Concentration of total proteins}$$

$$\text{Serum total proteins } (Y) = \frac{\text{OD of test}}{\text{OD of standard}} \times \text{Concentration of albumin}$$

Serum globulins = X – Y

$$\text{A/G ratio} = \frac{Y}{X-Y}$$

The values were expressed in g/100 ml.

Serum Glutamate Pyruvate Transaminase (SGPT)

In a test tube, 0.5ml buffered alanine 2-oxoglutarate substrate (pH 7.4) was taken and incubated at 37°C for 5 minutes. To this, 0.1 ml of test serum is added, mixed well and incubated at 37°C for 30 minutes. 0.5 ml of DNPH colour reagent is added to this mixture, mixed well and kept at room temperature for 20 minutes. To this mixture, 5.0 ml of NaOH (0.4N) is added, mixed well and kept at room temperature for 10 minutes. The OD of the mixture is read in a colorimeter at 505 nm against distilled water and refer to a standard graph with working pyruvate standard. The enzyme activity is expressed in units/ml.

Serum Glutamate Oxaloacetate Transaminase (SGOT)

The method of estimating SGOT is the same like that of SGPT estimation, but buffered L-aspartate 2-oxoglutarate substrate (pH 7.4) is to be used instead of SGPT substrate. The incubation period of the mixture is 60 minutes.

REVIEW QUESTIONS

1. How can you evaluate the toxicity of a pollutant in an aquatic organism?

2. How can you evaluate the toxicity of a pollutant in a terrestrial animal?

3. How can you determine the effect of temperature on toxicity of a pollutant?

4. How can you determine the effect of pH on toxicity of a pollutant?

5. Write down the method to determine the combined toxicity of pollutants.

6. Explain the qualitative methods to detect organochlorides, organophosphates and carbamates.

7. Explain the fingerprinting technique to analyse pesticide residues.

8. Explain the procedure to evaluate sublethal toxicity of a pollutant.

9. Write down the methodology of tissue preparation to observe histopathological alterations.

10. Explain the procedure of bone marrow and spleen smear.

11. How could oxygen consumption be determined in aquatic animals exposed to pollutants?

APPENDIX I

PREPARATION OF REAGENTS

Acrylamide Stock Solution

Acrylamide	30 g
Bis	0.8 g

Note: Distilled water was added to make the final volume 100 ml and stored in brown bottles upto 3 months at 4°C.

Alcoholic Bouin's Solution

Picric acid (alcholic, saturated)	75 ml
40% Formalin	20 ml
Glacial acetic acid	5 ml

Alcoholic Eosin

Eosin	1 g
70% Ethanol	100 ml

Alkaline Copper Reagent

Solution A

Sodium carbonate (anhydrous)	50 g
Sodium potassium tartarate (Rochelle salt)	50 g
Sodium bicarbonate	40 g

Sodium sulphate (anhydrous)	400 g
Distilled water	1600 ml
Note: diluted to 2000 ml	

Solution B

Copper sulphate	150 g
Distilled water	800 ml
Conc. H_2SO_4	0.5 ml
Note: diluted to 1000 ml	

Note: 96 ml of solution A was mixed with 4 ml of solution B.

Ammonium Persulphate Solution

APS	0.1 g
Distilled water	1 ml

Note: This solution was prepared freshly everyday.

Anthrone Reagent

Anthrone	50 mg
Thiourea	1 g
66%Conc. H_2SO_4	

Arsenomolybdate Colour Reagent

Solution A

Ammonium molybdate	100 mg
Distilled water	1800 ml
Conc. H_2SO4	84 ml

Solution B

Disodium orthoarsenite	12 mg
Distilled water	100 ml

Note: Solution A and B are to be mixed together and incubated at 37°C for 24–48 hrs.

Biuret Reagent

Solution A

Sodium citrate	17.3 g
Sodium carbonate	10.0 g
Distilled water	30 ml

Note: To be heated in water bath for 10–15 minutes.

Solution B

Copper sulphate	17.3 g
Distilled water	30 ml

Note: Solution A and B are to be mixed and made up to 100 ml with distilled water.

Borax Buffer (0.05 m: pH 9.2)

Borax	1.91g
Distilled water	100 ml

Bouin's Solution

Picric acid (aqueous, saturated)	75 ml
40% formalin	20 ml
Glacial acetic acid	5 ml

Ferric Chloride Solution (0.07)

Ferric chloride	6 g
Distilled water	100 ml

Gel Overlay Solution

Separating gel buffer	25 ml

10% SDS solution	1 ml
Distilled water	74 ml

Note: Final volume 100 ml

Glucose Solution

Glucose	100 mg
Saturated benzoic acid	100 ml

Glycine Buffer

Glycine	1.88 g
Distilled water	125 ml
0.1N NaOH	75 ml

Harris Haematoxylin

Haematoxylin	500 mg
Aluminium ammonium sulphate	20 mg
Mercuric oxide	500 mg

Note: This solution is to be boiled and cooled before mixing with Mercuric oxide. To be prepared one week prior to use.

Hydrochloric Acid (HCl)

1N Conc. HCl	8.62 ml

Note: Make up to 100 ml using distilled water.

6N Conc. HCl	51.72 ml

Note: Make up to 100 ml using distilled water.

Metaperiodate Solution

Ammonium acetate	77 g
Glacial acetic acid	60 ml

Na$_2$SO$_4$	650 mg
Distilled water	700 ml

Ninhydrin Reagent

Ninhydrin crystal	200 mg
n-butanol	100 ml

p-hydroxydiphenyl Reagent

p-hydroxydiphenyl phosphate	50 mg
3% NaOH	3 ml

p-nitrophenylphosphate Solution (PNP Solution)

2-*p*-iodophenyl-3-*p*-nitrophenyl-5-phenyl tetrazolium chloride	150 mg
Distilled water	25 ml

Phosphate Buffer (0.1m)

Solution A

Monobasic sodium phosphate	2.78 g
Distilled water	100 ml

Solution B

Dibasic sodium phosphate	5.37 g
Distilled water	100 ml

Note: For the desired range of pH, the following volumes of solution A and solution B were mixed and made up to 200 ml with distilled water.

Solution A (ml)	Solution B (ml)	pH
87.7	12.3	6.0
68.5	31.5	6.5

39.0	61.0	7.0
16.0	84.0	7.5
5.3	94.7	8.0

Potassium Chloride Solution

| KCl | 0.2g |
| Distilled water | 100ml |

Sample Buffer

Stacking gel buffer	25 ml
10% SDS Solution	4 ml
20% Glycerol	2 ml
0.02% Bromophenol blue	2 mg
0.2 M Dithiothreitol	0.31 g

Note: Distilled water was added to make the final volume 10 ml. Aliquots of 0.5 ml in eppendorf tubes were made and stored at $-20°$ C up to 6 months.

Potassium Phosphate Buffer (0.4 m, pH 7.2)

Solution A

| Potassium hydroxide | 2.98 g |
| Distilled water | 100 ml |

Solution B

| Potassium dihydrogen phosphate | 7.12 g |
| Distilled water | 100 ml |

Note: 35 ml of solution A mixed with 50 ml of solution B and made up to 100 ml with distilled water.

Saponifying Agent

Potassium hydroxide	50 g
Isopropanol	40 ml
Distilled water	600 ml

Separating Gel Buffer

Tris	18.15 g
Distilled water	75 ml

Note: pH was adjusted with 0.1 M HCl and the final volume was made to 100 ml with distilled water.

Sodium Dodecylsulphate Solution

SDS	1 g
Distilled water	10 ml

Note: To start with less volume of water was added and the final volume made to 10 ml.

Sodium Hydroxide (NaOH)

0.05N NaOH	200 mg
Distilled water	100 ml
0.1N NaOH	400 mg
Distilled water	100 ml
0.4N NaOH	1.6 g
Distilled water	100 ml

Sodium Succinate Solution (0.15 m, pH 7.0)

Sodium succinate	400 mg
Distilled water	100 ml

Stacking Gel Buffer

Tris	1.5 g
Distilled water	20 ml

Note: pH was adjusted with 1 M HCl and the final volume made to 25 ml with distilled water.

Tank Buffer

Tris	3.28 g
Glycine	14.413 g
SDS	1.0 g

Note: Distilled water was added to make the final volume to one litre and stored at room temperature for up to 1 month.

Zinc Chloride Reagent

Zinc chloride (anhydrous)	40 g
Glacial acetic acid	153 ml

Note: This solution was kept at 80°C for 2 hours 30 minutes.

APPENDIX II

IMPORTANT ABBREVIATIONS AND SYMBOLS USED IN TOXICOLOGY

ADI	Acceptable daily intake
BCF	Bioconcentration factor
BM	Biological monitoring
C	Central compartment
CP	Plasma concentration
Cl	Clearance
Cl_H	Hepatic clearance
Cl_H	Metabolic clearance
Cl_R	Renal clearance
Cl_S	Systemic clearance
EHC	Enterohepatic circulation
FRC	Functional residual capacity
GFR	Glomerular filtration rate
GSH	Glutathion (reduced)
GSSG	Glutathion (oxidized)
GST	Glutathion s-transferase
GT	Glucuronyl transferase
i.m.	intramuscularly
i.p.	intraperitonially
i.v.	intravenously

K	Elimination rate constant
K_A	Absorption rate constant
K_D	Dissociation constant
K_E	Excretion constant
K_M	Metabolic constant
MAC	Maximum allowable concentration
MFD	Mixed function oxidase
MRL	Maximum residue level
MTI	Mixture toxicity index
NOEL	No-observed effect level
P.O.	Oral (Per os)
ROS	Reactive oxygen species
RV	Residual volume
s.c.	Subcutaneously
STELs	Short-term exposure limits
T	Peripheral compartment
$t_{1/2}$	Elimination half-life
TLV	Threshold limit value
V_d	Volume distribution

GLOSSARY

Acceptable daily intake (ADI) The quantity of a substance that is ingested by the organism based on body weight over a lifetime without appreciable health risk.

Additivity The phenomenon where the combined effect of two toxic substances is equal to the sum of the effect of individual substances.

Aerosol A stable mixture of air and solid dust particles or droplets of fluid.

Aflatoxins A group of mycotoxins produced by *Aspergillus flavus* and other related moulds possessing carcinogenic and hepatotoxic properties.

Agonist Toxic substances which react with a receptor resulting in a biological effect through physiological and biochemical changes.

Alveolitis An inflammatory process in the alveoli characterized by the desquamation and proliferation of pneumocytes.

Anencephaly An anomaly in which there is complete absence of brain tissue with associated abnormalities in ears, eyes and neck.

Antagonism The phenomenon where toxic substances in a mixture counteract each other's action.

Antagonist The substance which reacts with a receptor resulting in the inhibition of the action of the agonist.

Atrophy The diminution in size and weight of a tissue or an organ due to decreased quantity of cytoplasm in cells and may be simple (shrinkage of individual cells) or numerical (reduction in number of cells).

Biological limit value (BLV) The value of the maximum permissible concentration of a toxic substance or its metabolite in biological systems.

Biomarker Measurable biochemical, physiological, cytological, immunological or molecular changes in biological systems which occur due to exposure to metabolites in the body fluids and are used to assess the risk of cancer prior to the onset of clinical symptoms of the disease.

Biotransformation system The collective processes by which the toxic substances are metabolized to form more water-soluble and readily excreted compounds.

Blood-brain barrier The covering on the capillaries in the brain and spinal cord that prevents the entry of toxic substances especially hydrophilic substances.

Carcinoma A malignant tumour in squamous, glandular or sensory epithelium, which is capable of infiltration and metastasizing.

Contaminants A number of non-natural and undesirable substances contained in foodstuff.

Control group A group of test organisms in an experiment, which is not exposed to the test solution under investigation.

Cytochrome P-450 A part of the mixed function oxidase (MFO) enzyme system that plays an important role in the oxidation of lipophilic compounds.

Cytotoxicology A sub-discipline of toxicology concerned with the study of cell toxicity and its mechanisms.

Dermatitis A disorder of skin due to cell proliferation, degeneration and inflammation.

Detoxification Process of production of metabolites, which are less toxic than the parent compound.

Dose The mass/quantity of a substance administered into the body of an organism expressed in absolute terms (mg, g, etc.) in relation to the organism's body weight.

Dysplasia The altered development of a particular region of the body due to the abnormality in size and shape of tissues.

Environmental compartments Air, water, soil and organisms constitute environmental compartments which are exposed to toxic substances.

Exposure chamber Inhalation toxicity tests in order to expose the test organisms to a test atmosphere under controlled condition.

Exposure phase Intoxication as a sequence of events starting from the exposure of an organism to a toxicant.

Foetotoxicity The deleterious effects such as lethality, growth impairment, physio-logical dysfunction, etc. in a foetus due to exposure to toxicants.

Graded toxic relation-ship Differences in the intensity of toxic response in relation to the dose of the toxic substance.

Half-life/Half-time The time interval during which the concentration of a toxicant is reduced by half.

Inflammation A local reaction from other cells and tissues leading to tissue damage.

Ion channels Specialized membrane proteins in the nerve membrane which produce an action potential through selective access of ions.

Isobole The line in a graph connecting points of equal strength.

Kinetics A mathematical study of the changes in the concentration of xenobiotics or their metabolites in the body of organisms after exposure.

Leukaemia A malignant disease in which there is abnormal production of blood cells in the bone marrow.

Local effects The injury or destruction of living cells at the site of first contact of the toxic substance in the organs of affected organisms.

Long-term exposure Continuous or repeated exposure of test organisms to a toxicant over a long period of time.

Lowest observed effect level (LOEL) The lowest concentration of a toxicant which will cause alterations in the body of target organisms when compared to control organisms of the same species under controlled exposure.

Lowest-observed-adverse-effect level (LOAEL) The concentration of a substance which will cause an adverse effect in target organisms when compared to control organisms of the same species under controlled conditions of exposure.

Lymphoma Malignant disease with abnormal cells in the lymphatic organs.

Maximum allowable concentration (MAC) The concentration of a toxic substance, which does not cause appreciable hazard on non-target organisms.

Maximum residue limit (MRL) The maximum concentration of any toxic residue recommended and permitted in or on foodstuff and animal feeds.

Metabolites The conversion products of a toxic substance that are more toxic or less toxic than the parent compound.

Metastasis The spreading of malignant tumour to distant regions from its site of origin.

Myelinopathy The alteration in the structure of neural membrane with the resultant loss of myelin sheath (demyelination).

Necrosis The morphological change in dying cells associated with major disturbances in both internal and external conditions brought out by physical (mechanical, electrical or thermal, ionizing and radioactive), chemical (toxins, pesticides, pharmaceutical chemicals, heavy metals), immunological (complement) and biological (viruses, bacteria, fungi, protozoa, metazoa) agents or by circulatory disturbances (ischaemia, hypoxia, anoxia).

Neuronopathy Lesions in the neuron due to degenerative changes in the cell body caused by toxicants.

Neuropathy Structural alterations of a part of the nervous system.

No-effect level (NOEL) The concentration or dose of a substance, which does not produce toxic effects in organisms on exposure.

Persistency The length of time at which a toxic substance remains in an environment, including within organisms, without being broken down.

Photo-allergic reaction The reaction produced at the site of skin on exposure to light where a hapten combines with the skin protein.

Phototoxicity The combined effect of the toxic substances and sunlight leading to disorders.

Potentiation The phenomenon where one substance in a mixture, which does not possess a toxic effect, would strengthen the action of another toxicant.

Quantal toxic relationship The dose-related differences in the intensity of toxic response in a natural population.

Receptor sites The binding sites in macromolecules to which toxicants bind more or less specifically causing a biological effect.

Risk assessment or risk analysis The process of characterization of the potential adverse health effects in humans on exposure to toxic chemicals.

Selective toxicity The ability of a toxicant to produce adverse effects only on one species.

Steatosis The fatty degeneration of the liver.

Synaptopathy A condition of impaired neurotransmission at the synaptic regions due to the action of xenobiotics.

Synergism The phenomenon where an inactive substance would enhance the action of another substance in a mixture.

Systemic effects The harmful effects caused by the toxicants on target organs in different parts of the body of organisms after absorption and distribution.

Target organs Specific organs in which the adverse effects of intoxication manifests.

Teratogenic substances Toxicants which cause damage in offsprings through parents in various ways.

Threshold dose A dose below which all organisms would remain alive and above which all would die.

Threshold limit value (TLV) The value of the maximum concentration of a toxic substance in the environment.

Toxicodynamic phase The sequence of events following the interaction of toxicants with target molecules resulting in harmful effect.

Toxicokinetic phase The whole process of entry, absorption, distribution, metabolism and excretion of toxicants.

Toxicodynamics The relationship between the concentration and effect of toxicants with specific emphasis on mechanism of action.

Xenobiotics The substances which are foreign to the body of organisms.

REFERENCES

Boyland, E. and Goulding, R. (1968). *Modern Trends in Toxicology.* Butterworths, London.

Finney, D.J. (1964). *Probit Analysis,* 2nd edn. Cambridge University Press, London.

Laws, E.A. (1981). *Aquatic Pollution,* John Wiley and Sons. New York.

Loomis, T.A. (1968). *Essentials of Toxicology.* Lea Febiger, Philadelphia.

Mehman, M.A. (ed.). (1976). *Advances in Modern Toxicology.* Hemisphere Publishing Co., USA.

Moriarity, F. (1983). *The Study of Pollutants in Ecosystem.* Academic Press, London.

Niesink, R.J.M., John de Vries and Hollinger, M.A. (eds.). (1996). *Toxicology Principles and Applications.* CRC Press, New York.

Paget, G.E. (1970). *Methods in Toxicology.* Blackwell Scientific Publishers, Oxford.

Rand, G.M. and Petrocelli, S.R. (eds.). (1985). *Fundamentals of Aquatic Toxicology.* Hemisphere Publishing Co., USA.

Wayland, J. and Hayes, J.R. (1975). *Toxicology of Pesticides.* Williams & Wilkins Company, USA.

INDEX

19488927R00198

Printed in Poland
by Amazon Fulfillment
Poland Sp. z o.o., Wrocław